Diversity and dynamics of the mammalian fauna in Denmark throughout the last glacial–interglacial cycle, 115–0 kyr BP

by

Kim Aaris-Sørensen

Acknowledgement

Financial support for the publication of this issue of Fossils and Strata was provided by the Carlsberg Foundation.

Contents

Diversity and dynamics of the mammalian fauna in Denmark throughout the last glacial–interglacial cycle, 115–0 kyr BP

KIM AARIS-SØRENSEN

Aaris-Sørensen, K. 2009: Diversity and dynamics of the mammalian fauna in Denmark throughout the last glacial–interglacial cycle, 115–0 kyr BP. *Fossils and Strata*, No. 57, pp. 1–59. ISSN 0024-1164.

This article presents new data on the changes in diversity and distribution in time and space of the mammalian fauna in Denmark throughout the Weichselian glaciation and the Holocene (ca. 115–0 kyr BP). Three different taphonomic pathways are recognized in the fossil assemblages. The oldest bone remains are found redeposited in glacial or glaciofluvial sediments belonging to the repeated advances and retreats of the Weichselian glaciers. The younger remains, which date to the time after the last glacial events (younger than ca. 17–16 kyr BP), are found *in situ* in lacustrine, fluviatile or marine sediments *either* in natural deposits *or* in man-made refuse deposits. In all, 77 terrestrial and marine mammal species have been identified and described in detail as regards first and last appearance data (FAD and LAD), number of dated remains (NDR) and the inferred time range in the Danish/south Scandinavian area. Analyses of the data show that the central European mammoth steppe fauna expanded northward into southern Scandinavia during all Weichselian interstadials. The dynamics of the fauna match the advances and retreats of the ice cap and are best explained as a combination of expansions and local extinctions of marginal populations. After the Last Glacial Maximum (LGM) a dramatic increase in species richness is seen between 15 and 11 cal. kyr BP followed by a decline towards a more moderate level between 11 and 9 kyr and a steady state in the last 9000 years with a mean richness of ca. 30 species. Two opposing processes, colonizations and extinctions, regulate the diversity and this also caused large changes in species composition with faunal turnover rates of 90% to 60% during the first 4000 years and 20% to 4% during the last 6000 years. These changes are analysed in relation to the climate-induced environmental changes and increasing human impact. It is shown that the expansion of herbivore populations was controlled by the rate of 'habitat migration' – the time lag in vegetational response to the climatic improvement and furthermore that the Younger Dryas setback induced a local extinction of several species around 12.6 cal. kyr BP. The large number of species (38) forming the peak during the 15 to 11 kyr interval was a mixture of species today considered as ecologically incompatible. The assemblages are referred to as a non-analogue or disharmonious fauna and it is explained as the result of individualistic responses by the species in accordance with their individual tolerance limits during environmental changes. Man-made habitat fragmentations after the introduction of agriculture and husbandry around 6000 BP and a steady increase in encounters with and persecution by a growing human population led to extinctions. At the same time, however, new species were attracted by the new open cultural landscape and by human habitations and activities. The transformation of continental Denmark into a group of islands and peninsulas around 8 cal. kyr BP led to further habitat fragmentations and to isolated and vulnerable island populations. An impoverished fauna is documented on the island of Sjælland which experienced a local extinction of four carnivores and two ungulates during the interval 8 to 7 kyr. The marine mammals are represented by 15 whale species, four true seals, the walrus and the polar bear. The oldest remains belong to cold-adapted species which were present in northern Denmark whenever an arctic–subarctic palaeo-Kattegat–Skagerrak existed during the Weichselian. The youngest, on the other hand, include temperate- and warm-adapted species and they have been spread more widely across the country and much further south in accordance with the Mid–Late Holocene transgressions creating the Inner Danish Waters. □ *Denmark, diversity, dynamics, Holocene, Mammalia, Weichselian.*

Kim Aaris-Sørensen [kaaris@snm.ku.dk], Zoological Museum, Natural History Museum of Denmark, University of Copenhagen, Universitetsparken 15, DK-2100 Copenhagen Ø, Denmark.

Introduction

This paper summarizes and scientifically documents the history of the mammalian fauna in Denmark throughout the last glacial–interglacial cycle. So far a comprehensive faunal history survey has only been given in popular or semi-popular publications as recently by Aaris-Sørensen (1998, 2007), whereas scientific contributions have been confined to single taxonomic groups (e.g. Carnivora, Degerbøl 1933; Pinnipedia, Møhl 1971a), single species (e.g. *Ursus maritimus*, Nordmann & Degerbøl 1930; Aaris-Sørensen & Petersen 1984; *Saiga tatarica*, Degerbøl 1932; Aaris-Sørensen *et al.* 1999; *Bos primigenius*, Degerbøl & Fredskild 1970; Aaris-Sørensen 1999; *Rangifer tarandus*, Degerbøl & Krog 1959; Aaris-Sørensen *et al.* 2007; *Megaloceros giganteus*, Aaris-Sørensen & Liljegren 2004; *Mammuthus primigenius*, Aaris-Sørensen *et al.* 1990) or to restricted periods (e.g. late Middle Weichselian, Aaris-Sørensen 2006; Late Weichselian–Early Holocene, Aaris-Sørensen 1992; the Atlantic period, Aaris-Sørensen 1980).

Because of its highly dynamic glacial history followed by a likewise highly changeable Post Glacial period, the Danish/south Scandinavian area presents a very interesting area for studying the interrelationship between the environmental changes (as observed by geologists, climatologists, palaeobotanists and archaeologist) and the faunal changes.

The aim of this article is first of all to let the mammals speak by describing the diversity and the dynamics as precisely as the data allow and then to compare and discuss the observed patterns with the scenarios set up by combined geological, climatological, botanical and archaeological investigations.

Last glacial–interglacial cycle

Chronologically this work includes the last cold stage, the Weichselian Glacial from ca. 115 to 11.7 kyr BP and the present temperate 'interglacial' stage, the Holocene which covers the last 11.7 kyr. The highly changeable Danish–south Scandinavian environment during this last glacial–interglacial cycle is outlined in Figure 1. The alternating stadials and interstadials and the environmental changes involved are linked to the deep-sea oxygen isotope record which can be seen as a proxy for global land ice cover and temperature (a

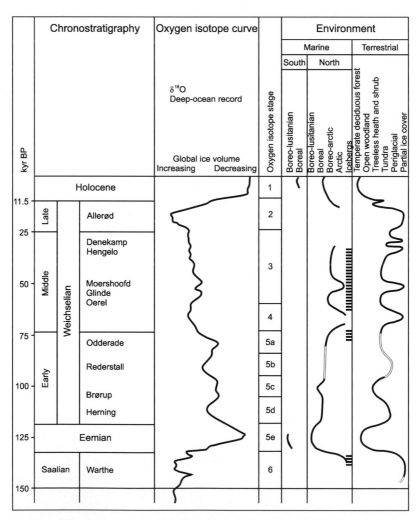

Fig. 1. Climatic and environmental changes in Late Pleistocene Denmark. (After Houmark-Nielsen *et al.* 2006).

temperature curve for the last 15 kyr is shown in Fig. 32, see p. 32). Figure 1 clearly illustrates the extreme climate fluctuations and recurring changes in the marine and terrestrial environments of the period.

Despite the difficulties and uncertainties involved in dating the stadials and interstadials and in establishing the extent of the ice sheet during these periods it can be firmly concluded that ice-free conditions prevailed in the Danish area during the Weichselian (Houmark-Nielsen & Kjær 2003; Houmark-Nielsen 2004; Houmark-Nielsen *et al.* 2005, 2006).

The repeated growth and contraction of the ice sheet controlled the change in land/sea/ice configurations throughout the period. Eustatic sea-level changes and isostatic depression of the land surface followed by postglacial rebound determined the palaeogeography. A series of selected palaeogeographical reconstructions are presented in Figure 2 showing the most important ice advances and ice-free interstadials of the Weichselian as well as the final deglaciation.

In the Early Weichselian (Marine Isotope Stage (MIS) 5d–5a) no ice advances reached Denmark (Fig. 2A). During the Middle Weichselian (MIS 4–3), on the other hand, the global ice volume was much larger and three ice advances have been recorded from the Danish area. The first one came from the north around 65 kyr BP to reach as far south as northern Jylland and Kattegat and the second moved through the Baltic depression and reached the south-eastern parts of Denmark around 50 kyr BP (Fig. 2B); finally, a third short-lived glacier reached the south-eastern-most parts around 35 kyr BP (Fig. 2D), once more through the Baltic depression. The Scandinavian Ice Sheet advanced south to its maximum extent during the beginning of the Late Weichselian (MIS 2) when the global ice volume was at its largest. The ice reached its maximum position in the Danish region around 22 kyr BP following the so-called Main Stationary Line in Mid-Jylland (Fig. 2F). The final deglaciation of the ice sheet started as early as around 19 kyr BP in the Danish region and after several still-stands and readvances of the ice front the Danish area was finally free of ice around 16 kyr BP (Fig. 2G, H).

Based on studies of mollusc assemblages, changes in the marine environment have been recorded covering boreal-lusitanian to arctic conditions (Petersen 2004; Fig. 36, see p. 49). The terrestrial environment varies between harsh periglacial conditions close to the ice margins during the coldest stadials and open, treeless vegetation dominated by grasses and sedges with scattered shrubs and dwarf shrubs during the warmer ice-free interstadials, or even open woodland with birch and pine during the mildest periods. Vegetational development during the Holocene is shown in Figure 3. It appears that the postglacial warming gave rise to an open birch–pine forest followed by a

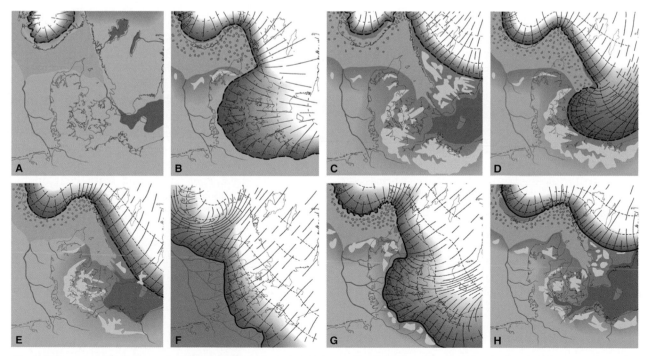

Fig. 2. Palaeogeographical maps of southern Scandinavia. A, 100–90 kyr BP; B, 50–45 kyr BP; C, 45–35 kyr BP; D, 35–33 kyr BP; E, 33–31 kyr BP; F, 23–21 kyr BP; G, 19–18 kyr BP; H, 16–14.5 kyr BP. Grey shaded land areas are dominated by dead-ice. (After Houmark-Nielsen *et al.* 2005).

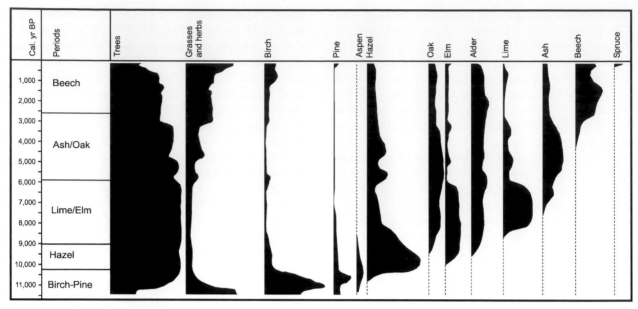

Fig. 3. Generalized pollen diagram showing forest development in Denmark during the Holocene. (After Odgaard 2006).

more dense hazel-dominated woodland which gradually transforms into a dense, mixed deciduous forest dominated by first lime/elm, then ash/oak and finally by beech. A close look at the pollen diagram reveals a general decline in the tree pollen percentage and a corresponding increase in the combined grass/herb pollen percentage beginning around 6000 BP. This progressive deforestation is a result of the introduction of agriculture and husbandry in Denmark and the Neolithic farmers' need of pastures and arable land.

The arrival of the first humans as far north as Denmark and south Scandinavia took place as late as at the end of the Bølling period, i.e. around 14.2–14.0 cal. kyr BP (Mortensen 2007; Mortensen *et al.* 2008). These were the Late Palaeolithic hunter-gatherers of the Havelte stage of the Hamburgian culture. The classic Hamburgian has its main distribution in northernmost Germany and Holland, with an outlier in western Poland (Tromnau 1975), whereas the Havelte is confined to northern Holland, northern Germany and the southern part of Denmark.

The next cultural group to arrive was the Federmesser or the Arched Backed Points (Schild 1984). This particular group is only represented by a couple of sites in southern Jylland (Fischer1990; Holm 1993) and eastern Denmark (Petersen 2001, 2006) and none of these have been radiocarbon dated. Based on the evidence from northern Europe, these sites are probably to be dated to the period between 14.0 and 13.5 cal. kyr BP, corresponding to the Older Dryas as well as the beginning of the Allerød warming period.

The third human group to occupy the present area of Denmark was the Brommian, and sites of this character are found from all over the south Scandinavian area including Skåne. As the few radiocarbon dates centre around 13.0 cal. kyr BP, the Brommian is found during the Allerød period and perhaps also during the first part of the Younger Dryas (Petersen in press).

The fourth human group to have a sporadic presence across the area was the Ahrensburgian. Sites of this type are found mainly to the south of the Danish/German border and so far only a few have been found in the Danish area (Petersen & Johansen 1993; Møbjerg & Rostholm 2006). According to one radiocarbon date from Schleswig, giving an age of ca. 13.0 cal. kyr BP (Clausen 1996), this group might even overlap with the Brommian, whereas the rest of the dates correspond to the end of the Younger Dryas.

The first Mesolithic culture, the Maglemosian, seems to correspond to the beginning of the Preboreal period (Hansen *et al.* 2004) and from then on the Danish area was continuously occupied by humans. The Maglemosian, so far mostly an inland culture, ends around 8.5 cal. kyr BP, when the Kongemosian takes over until 7.5 cal. kyr BP (Sørensen 1996).

The Neolithic started around 6.0 cal. kyr BP with the Trichterbecher culture, but during the first 400 years it was as much a hunter-gatherer as a farming existence (Andersen 2008; Petersen & Egeberg in press). From 5.6 cal. kyr BP it had a fully developed agricultural subsistence base. This continued through the Bronze Age, 4.0–2.5 cal. kyr BP and the Iron Age, 2.5–1.2 cal. kyr BP (AD 800). The first towns (e.g. Ribe) develop around AD 800.

Material and taphonomy

This investigation is based on a very large number of subfossil bone remains representing 77 mammalian species. They make up the greater part of the national Danish collection of Late Pleistocene and Holocene vertebrates which is housed at the Zoological Museum (ZMUC) (part of National History Museum of Denmark, SNM) at the University of Copenhagen. Relevant finds from adjacent areas (Norway, Sweden, Britain and the northern European lowlands) are included as additional evidence in the discussion of the history of many of the species.

The remains represent extant species as well as species now locally or even globally extinct. Nevertheless, acknowledging the fundamental problem associated with 'negative evidence' (absence of evidence is no evidence of absence) biological evidence such as the modern ecology and geographic distribution of the species is also taken into consideration. This means that species now living in the area but which totally lack a subfossil record might be included and that the time range of a species with an obvious incomplete record might be adjusted.

Three main categories can be recognized in the subfossil collection. The oldest (>ca. 20 ^{14}C kyr BP) consists of bone remains found redeposited in glacial or glaciofluviale sediments belonging to the repeated advances and retreats of the Weichselian glaciers. They are normally found along exposed cliffs or in gravel pits as single bones or teeth of large mammals, rarely complete and all showing more or less clear signs of having been transported by ice or water (Fig. 4).

These specimens make it clear that only the most resistant elements can survive the glacial events and, moreover, that they can only have been transported a short distance before redeposition. This is important as it corroborates the assumption that they do represent a local fauna.

The two other main categories both belong to the time after the last glacial events (<ca. 17–16 cal. kyr BP). These remains are found in lacustrine, fluviatile or marine sediments either in natural deposits or in man-made refuse deposits. Bone assemblages representing rather large catchment areas may be found among the former, but the most conspicuous ones consist of more or less complete single skeletons of naturally trapped larger animals. As a result of dead ice formation during the deglaciation period large parts of Denmark were densely covered by kettle holes. Thousands of animals drowned in these lakes which later grew into bogs. The animal remains have been found during peat cutting and drainage operations embedded in the gyttja underlying the peat (Figs 5, 6).

More or less complete skeletons may also be found in the man-made middens investigated by archaeological excavations. However, the typical bone assemblage found in an archaeological context consists of thousands of highly fragmented and worked bones (Fig. 7). Although these only represent certain selected parts of the fauna zooarchaeological materials,

Fig. 4. Skull of musk ox (*Ovibos moschatus*) rounded and polished after being transported by water. Bannebjerg near Helsinge, N Sjælland. Age 28.5 kyr BP. (Photo: Geert Brovad).

Fig. 5. Complete skeleton of reindeer (*Rangifer tarandus*) found at Villestofte near Odense, Fyn. Age: ca. 14.0 cal. kyr BP. (Photo: Geert Brovad).

Fig. 6. Part of the Danish aurochs (*Bos primigenius*) collection at the Zoological Museum, University of Copenhagen. Time range: ca. 11.4–3.0 cal. kyr BP. (Photo: Geert Brovad).

nevertheless contribute considerably to faunal history, especially by adding the medium-sized mammals which rarely drown and are seldom retrieved in peat bogs. Furthermore, modern excavation techniques, applied for the last 30 years, which includes water screening of selected samples or even of the entire deposit, have also substantially increased the amount of small mammal bones (Aaris-Sørensen & Andreasen 1995).

Dating and chronology

Only well-dated specimens are included in this investigation with direct radiocarbon dating of bone collagen being the most used and most important method. However, before the radiocarbon dating method was introduced and even afterwards when it was not yet common procedure, direct dating by pollen analysis also played a major role. A large number of large mammals in the ZMUC collection discovered in connection with peat cutting during and immediately after the Second World War have been pollen dated. Many of these specimens have later been re-dated by radiocarbon analysis and the two sets of results turn out (in most cases) to be in good agreement with each other. Therefore, high-quality pollen dating with unambiguous results, especially if linked to the more changeable parts of the Late and Post Glacial vegetational history, has been 'translated' into reliable intervals on a calendar time-scale.

Besides, also a large number of indirectly dated specimens are included. In cases with a clear and well-

described stratigraphy, datings of other bones or other organic remains found in the same layer have been applied to the specimens under investigation. Indirectly dated remains most often come from archaeological sites and here stratigraphy and chronology are normally cross-checked by applying several dating methods, most commonly pollen analysis, radiocarbon dating and also a typologically based relative dating of the artefacts. Odd and unexpected occurrences have been checked by radiocarbon dating.

The fundamental differences in taphonomy and number of specimens between the redeposited remains of the Weichselian megafauna and the *in situ*

Fig. 7. Typical example of a mammalian bone assemblage from an archaeological site. In this case it contains limb bone fragments of aurochs and red deer. Age: Maglemosian culture, ca. 9.5 cal. kyr BP. Scale bar = 10 cm. (Photo: Geert Brovad).

recovered remains from the Late and Post Glacial periods require two different ways of presenting their chronological distribution.

For the Weichselian, up to the onset of the final deglaciation, a regional chronostratigraphy is used with stadials and interstadials linked to a time-scale which is partly based on OSL datings and radiocarbon datings (Fig. 25) (Houmark-Nielsen & Kjær 2003; Houmark-Nielsen 2004). Individual dates of mammals are given in uncalibrated ^{14}C yr BP.

The last 17 000 years, covering what is traditionally named the Late and Post Glacial, have been divided into a series of marked climatic events according to the isotopic Greenland ice-core record (Fig. 31, see p. 42). The time-scale follows a new Greenland ice core chronology common to the DYE-3, GRIP and NGRIP ice cores, named the Greenland Ice Core Chronology 2005 (GICC05) (Rasmussen *et al.* 2006). The isotope event names used refer to those of Björck *et al.* (1998). Although originally defined as biostratigraphic units the terms Oldest Dryas, Bølling, Allerød, etc. are nevertheless used in the text throughout this work to name these climate periods. The time range of the individual mammalian species is plotted against this calendar time-scale by calibrating the radiocarbon dates using the OxCal program, 4.0 version (online) (Bronk Ramsey 1995, 2001). Calibrations are with 1 SD. Besides the calibrated date, designated as 'cal. kyr BP', the FADs and LADs (see next section) are also shown as the original dating results in conventional ^{14}C yr BP followed by the radiocarbon laboratory number.

Systematic section

This section summarizes all available data concerning the history of all native species. Denmark is treated as a unified whole despite the fact that it was transformed into a group of islands and one big peninsula some 8000 years ago. The regional differences in the mammalian faunas resulting from this event are discussed separately in the section on Islands. As subfossil records of bats (Chiroptera) from Denmark are almost non-existent (Baagøe 2007, p. 57) this group has not been included. Historical data and remarks on possible first appearance of some of the 17 bat species occurring in present-day Denmark can, however, be found in Baagøe (2001).

The primary evidence presented for each species consists of the first and last appearance data (FAD and LAD) and the number of dated records available per species (NDR). The latter is a simple qualitative record, counting the occurrence of a species at a certain site at a certain time as one (1) despite the number of specimens identified in the given assemblage.

If the dated records are scarce and/or discontinuous the primary evidence from Denmark is in many cases combined with additional data from adjacent areas in order to give a more probable time range. In some cases even actualistic evidence is taken into account, including facts from the modern ecology and geographical distribution of the species.

As it seems appropriate to gather these arguments species by species this section will inevitably anticipate the more general discussions in later sections. In order to facilitate these discussions the thousands of records presented in the Systematic section are compiled in tables and charts. The few Weichselian records dating from before the Last Glacial Maximum (Pre-LGM) are listed and plotted individually (see the section on 'The terrestrial fauna prior to LGM). LGM is here defined as the time of the maximum distribution of the Scandinavian Ice Sheet in Denmark about 22 cal. kyr BP when it reached the Main Stationary Line (Fig. 2F). The many Post-LGM records of terrestrial mammals are shown in a combined species range chart (see the section on 'The terrestrial fauna after LGM'). On this chart solid chrono-lines based on subfossil evidence are easily distinguished from dotted lines based partly or entirely on actualistic evidence. Finally, the marine mammal record is plotted against the different stages of marine environments of the Danish waters throughout the Weichselian and Holocene (see the section on The marine fauna). An approximate location of the finds mentioned in the text can be found with the help of the map in Figure 8.

Insectivora

Erinaceus europaeus

Hedgehog

First appearance datum. – Mesolithic (Maglemosian) sites of Lundby, Mullerup, Sværdborg and Holmegård, ca. 9.5 cal. kyr BP (Winge 1903, 1919, 1924; Aaris-Sørensen 1976; Rosenlund 1980).

Remains of *E. europaeus* are found almost exclusively in archaeological contexts. As all the Danish Mesolithic sites from the Preboreal lack any faunal remains, the FAD of many, especially smaller, mammals, has automatically been linked to the classical Maglemosian sites of the Boreal period. From an ecological point of view, however, an earlier immigration would be expected for many of these species.

Fig. 8. Location map of Denmark with names mentioned in the text. 1 = Hirtshals, 2 = Nørre Lyngby, 3 = Hjørring, 4 = Frederikshavn, 5 = Sæby, 6 = Thisted, 7 = Brovst, 8 = Nørre Sundby, 9 = Løgstør, 10 = Ålborg, 11 = Skive, 12 = Randers, 13 = Grenå, 14 = Skjern, 15 = Ikast, 16 = Silkeborg, 17 = Århus, 18 = Ebeltoft, 19 = Horsens, 20 = Vejle, 21 = Kolding, 22 = Ribe, 23 = Vojens, 24 = Haderslev, 25 = Tønder, 26 = Bogense, 27 = Odense, 28 = Assens, 29 = Fåborg, 30 = Fåreveile, 31 = Holbæk, 32 = Gilleleje, 33 = Helsinge, 34 = Helsingør, 35 = Hillerød, 36 = Frederikssund, 37 = Allerød, 38 = Vedbæk, 39 = Copenhagen, 40 = Dragør, 41 = Mullerup, 42 = Bromme, 43 = Sorø, 44 = Ringsted, 45 = Skælskør, 46 = Holmegård, 47 = Næstved, 48 = Lundby, 49 = Sværdborg, 50 = Nakskov, 51 = Nykøbing Falster, 52 = Tågerup, 53 = Lockarp, 54 = Arrie, 55 = Hässleberga.

Last appearance datum. – Extant.

Number of dated records. – 44, continuously dispersed between FAD and LAD.

Adjacent areas. – A first appearance prior to 9.5 cal. kyr BP is supported by findings in adjacent areas. Gramsch (2000) reports hedgehog remains from the site of Friesack in Brandenburg dated to ca. 10.3–9.6 cal. kyr BP (Zeitstufe III) and Fraser & King (1954) from the site of Star Carr in Yorkshire ca. 11.0–10.6 cal. kyr BP. These sites are only about 150 km further south than the Danish Maglemosian sites; at another 150 km further south, King (1962) reports hedgehog bones from Berkshire at the site of Thatcham dated as early as ca. 11.5–11.0 cal. kyr BP. These earlier occurrences close to the Danish area together with direct radiocarbon dates of first Holocene appearances of species such as *Meles meles*, *Equus ferus*, *Alces alces*, *B. primigenius* and *Bison bonasus* between 11.6

and 11.2 cal. kyr BP (see later this section) point at a similar Early Preboreal immigration of *E. europaeus*.

Inferred time range. – ca. 11.4–0 cal. kyr BP.

Sorex araneus

Common shrew

First appearance datum. – Four specimens recovered at a Late Weichselian freshwater bed at Nørre Lyngby, northern Jylland and dated to between 11 590 ± 130 and 11 120 ± 160 ^{14}C yr BP (AAR-1509–11 and 1908–10) (Aaris-Sørensen 1995) corresponding to ca. 13.6–13.1 cal. kyr BP (Fig. 9).

Last appearance datum. – Extant.

Number of dated records. – 10, discontinuously dispersed between FAD and LAD. The few dated

Fig. 9. Two soricid mandibles recovered at the Late Weichselian freshwater bed at Nørre Lyngby, N Jylland. The smaller mandible (top) belongs to the pygmy shrew (*Sorex minutus*) and the larger (bottom) belongs to the common shrew (*Sorex araneus*). Age: ca. 13.6–13.1 cal. kyr BP. (Photo: Geert Brovad).

records of *S. araneus* show a considerable time gap between the first appearances in the middle Allerød and the next following as late as in the middle of the Atlantic period at Late Mesolithic Ertebølle sites (Aaris-Sørensen & Andreasen 1995). The modern ecological requirements of the species are in agreement with the Late Glacial FAD during the warmer Allerød interstadial and also with the species being continuously present throughout the Post Glacial. The absence of evidence in Preboreal reflects the lack of faunal remains on Danish archaeological sites from this period and continuing absence in Boreal is probably a result of the excavation techniques applied in the early and middle 1900s when the classical Danish Maglemosian sites were investigated. At that time, water screening had not yet become a natural part of the excavation procedure; so, almost all small mammal bones were overlooked.

At the end of the Late Glacial, the Younger Dryas cooling turned the open Allerød woodland into a grass and shrub tundra again. There are indications in the record of some of the large mammals (see e.g. *Castor fiber*, *Ursus arctos*, *A. alces*, *M. giganteus* and *E. ferus*) that this short but severe cooling led to a local extinction, a Younger Dryas Induced Pause lasting from ca. 12.6 until a re-immigration took place (without *Megaloceros*) at the beginning of Holocene. This scenario might also have included *Sorex araneus*.

Inferred time range. – ca. 14.0–0 cal. kyr BP with a probable Younger Dryas Induced Pause ca. 12.6–11.4 cal. kyr BP.

Sorex minutus

Pygmy shrew

First appearance datum. – A single specimen (mandible) recovered at a Late Weichselian freshwater bed at Nørre Lyngby, northern Jylland and dated to between 11 590 ± 130 and 11 120 ± 160 ^{14}C yr BP (AAR-1509–11 and 1908–10) (Aaris-Sørensen 1995) corresponding to ca. 13.6–13.1 cal. kyr BP (Fig. 9).

Last appearance datum. – Extant.

Number of dated records. – 3. These three dated records of *S. minutus* show a considerable time gap between the first appearance in the middle of the Allerød, and the next following as late as in the middle of the Atlantic period at a Late Mesolithic Ertebølle site, Bjørnsholm south of Løgstør in northern Jylland (Andersen 1993; small mammals identified by E. O. Heiberg, unpublished ZMUC files) and in the Early Subboreal at a Neolithic site, Hov near Thisted in northern Jylland. Further considerations regarding the actual time range are identical to those given above in the case of *S. araneus*.

Inferred time range. – ca. 14.0–0 cal. kyr BP with a probable Younger Dryas Induced Pause ca. 12.6–11.4 cal. kyr BP.

Neomys fodiens

Water shrew

First appearance datum. – Two records are known from two Mesolithic Ertebølle sites from Middle/Late Atlantic, Lystrup Enge, north of Århus (identified by Liv Ljungar, unpublished ZMUC files) and Maglemosegård, Vedbæk, northern Sjælland (Aaris-Sørensen & Andreasen 1995).

Last appearance datum. – Extant.

Number of dated records. – 3. Besides the two Mesolithic specimens the species is only recorded from Storring west of Århus, dated to the 1600s.

Recent evidence. – Despite the lack of primary evidence, the modern geographical distribution and ecological requirements of the species make it obvious to assume a Late Glacial immigration together with *S. araneus* and *S. minutus* followed by a continuous presence until today, although with a possible Younger Dryas Induced Pause (for further discussion, see *S. araneus*).

Inferred time range. – ca. 14.0–0 cal. kyr BP with a probable Younger Dryas Induced Pause ca. 12.6–11.4 cal. kyr BP.

Desmana moschata
Russian desman

First and last appearance data. – Remains of *D. moschata* have been found three times at the Late Weichselian freshwater bed at Nørre Lyngby, northern Jylland. Fifty years separate the first and second discovery and 20 years the second and third. All specimens were recovered at the coastal cliff where the freshwater bed is exposed by the sea through an annual erosion of 1–2 m. The first discovery, an almost complete skull with mandibles, has been indirectly radiocarbon dated to 11 430 ± 160 ^{14}C yr BP (K-2900) (Bondesen & Lykke-Andersen 1978) (Fig. 10), the second, a skull fragment, comes from layers ascribed to the Younger Dryas (Enghoff 1984) and the third, 10 different bones, has been indirectly radiocarbon dated to between 11 590 ± 130 and 11 120 ± 160 ^{14}C yr BP (AAR-1509–11 and 1908–10) (Aaris-Sørensen 1995) corresponding to ca. 13.6–13.1 cal. kyr BP.

Number of dated records. – 3.

Adjacent areas. – The spread of a dry, continental steppe environment into Europe during the Weichselian was followed by an expansion of steppe elements such as *D. moschata* (see also *Ochotona pusilla*, *Spermophilus major*, *Microtus gregalis* and *S. tatarica*). The maximum distribution towards north-west was reached during the Late Glacial and in addition to the Danish specimens there are records from the Late Glacial sites of Stellmoor and Meindorf in Holstein (Krause & Kollau 1943; Benecke 2004).

Inferred time range. – ca. 14.0–11.5 cal. kyr BP.

50 mm

Fig. 10. An almost complete skull with mandibles of a Russian desman (*Desmana moschata*) found at the Late Weichselian freshwater bed at Nørre Lyngby, N Jylland. Age: ca. 13.4–13.1 cal. kyr BP. (Photo: Geert Brovad).

Talpa europaea
Mole

First appearance datum. – First record from a Mesolithic site, Maglemosegårds Vænge, Vedbæk, northern Sjælland dated to Middle Atlantic (layer 5: 6120 ± 100 ^{14}C yr BP (K-3171) (Christensen 1982; Aaris-Sørensen & Andreasen 1995).

Last appearance datum. – Extant.

Number of dated records. – 10. Except for a gap between Early and Late Subboreal the 10 dates are evenly dispersed between FAD and LAD. Concerning the lack of Danish records from Preboreal and Boreal – see the discussion on *S. araneus*.

The fossorial habits exhibited by many small mammals and especially by *T. europaea* and *Arvicola terrestris* have often cast doubt on whether they were actually contemporary with the deposit in which they were recovered. However, recent intruders are often found more or less complete, still articulated and with whitish or light yellow coloured bones (and in the case of *A. terrestris*, e.g. still with the characteristic yellow-orange enamel colour of the incisors). The specimens included in this investigation are all found as single bones and present the same colour and patina (often light to dark brownish) as the rest of the bones in the deposit. They are therefore believed to be true members of the assemblage in which they were recovered.

Adjacent areas. – Early Preboreal records are known from the Mesolithic site of Bedburg-Königshoven in the Lower Rhineland (Behling & Street 1999; Street 1999) and from Thatcham in Berkshire (King 1962).

Recent evidence. – Today the distribution of *T. europaea* seems to be limited by the presence of a soil and by the abundance and all-year round accessibility of earthworms. It has a southern distribution in Scandinavia covering only Denmark and southern Sweden; further east it includes the Baltic countries and only the southern part of Finland. It is unlikely that a suitable soil with earthworms and other invertebrates sufficiently abundant could have been developed in Denmark until the beginning of the Preboreal. The late Middle Atlantic FAD is contradicted by finds of oak fern (*Gymnocarpium dryopteris*), a brown earth-associated plant, which has been recorded in a pollen sample from Hassing Huse Mose with a date of around 11.5 cal. kyr BP (Andersen 1995).

Inferred time range. – ca. 11.4–0 cal. kyr BP.

Lagomorpha

Lepus europaeus

Brown hare

First appearance datum. – Remains from several Early Neolithic sites in Jylland as well as the islands of Sjælland and Langeland (e.g. Kolind, Sølager, Årby, Gadegård and Lindø) dated around 5.5–4.5 cal. kyr BP.

Last appearance datum. – Extant.

Number of dated records. – 58, continuously dispersed between FAD and LAD.

Inferred time range. – ca. 5.5–0 cal. kyr BP.

Lepus timidus

Mountain hare

First and last appearance data. – Only two records of this species, both directly radiocarbon dated. The oldest was found in Køge Bugt (off Solrød Strand) and dated to 12 190 ± 90 (AAR-4177) corresponding to ca. 14.1–13.9 cal. kyr BP. The other and younger specimen was found at Overgaards Mergelleje, Vejle and dated to 9910 ± 65 (AAR-4176) corresponding to ca. 11.4–11.2 cal. kyr BP.

Number of dated records. – 2.

Adjacent areas. – The mountain hare is well known from the last glacial period all over Europe (see Thulin & Flux 2003). In line with the Danish finds are a few remains reported from two kettle holes at Hässleberga in south-western Skåne. The bones were found together with numerous reindeer bones in a calcareous detritus gyttja of Late Glacial origin (Larsson *et al.* 2002).

Inferred time range. – Pre-LGM: Early and Middle Weichselian interstadials. Post-LGM: ca. 14.5–9.5 cal. kyr BP.

Ochotona cf. pusilla

Steppe pika

First and last appearance data. – Two bones of a pika, the characteristic squamous temporal bone and a fragment of a mandible with M_3, have been recovered at a Late Weichselian freshwater bed at Nørre Lyngby, northern Jylland and dated to between 11 590 ± 130 and 11 120 ± 160 ^{14}C yr BP (AAR-1509–11 and 1908–10) (Aaris-Sørensen 1995) corresponding to ca. 13.6–13.1 cal. kyr BP. More specimens are needed for a satisfactory species identification. Based on both ecological and morphological grounds, however, the extant steppe pika, *O. pusilla* seems the most likely choice and so far all *Ochotona* remains found in Western Europe have been assigned to this species.

Number of dated records. – 2.

Adjacent areas. – The spread of a dry, continental steppe environment into Europe during the Weichselian was followed by an expansion of steppe elements as *O. pusilla* (see also *D. moschata*, *S. major*, *M. gregalis* and *S. tatarica*). The steppe pika reached its maximum north-western extent during the Late Glacial and besides the Danish specimens is recorded from, e.g. southern Britain (Yalden 1999; Fisher & Yalden 2004), the Rhineland (Street & Baales 1999) and south-east Belgium (Cordy 1991) where it faced a local extinction during the Early Preboreal.

Inferred time range. – ca. 14.0–11.5 cal. kyr BP.

Rodentia

Sciurus vulgaris

Red squirrel

First appearance datum. – Mesolithic (Maglemosian) sites of Mullerup and Sværdborg, ca. 9.5 cal. kyr BP (Winge 1903; Aaris-Sørensen 1976).

Last appearance datum. – Extant.

Number of dated records. – 34, continuously dispersed between FAD and LAD, although with a gap in Late Subboreal between ca. 4.5 and 2.5 cal. kyr BP. For reasons discussed in the case of *E. europaeus* the lack of Preboreal bone remains cannot be taken as reliable evidence of absence of the species during that period.

Adjacent areas. – The only other Early Holocene record of *S. vulgaris* from northern Europe comes from the site of Friesack in Brandenburg dated to ca. 10.3–9.6 cal. kyr BP (Zeitstube III) (Gramsch 2000).

Recent evidence. – The modern distribution of the species covers all forested parts of the Palaearctic region where tree seeds from both broad-leaved and conifer species are the main food resource. These ecological requirements make the earliest immigration into southern Scandinavia probably between ca. 11.3 and 10.8 cal. kyr BP when a light birch–pine forest was established in the area (Mortensen 2007).

Inferred time range. – ca. 11.2–0 cal. kyr BP.

Spermophilus major

Russet ground squirrel

First and last appearance data. – Remains of *S. major* (the taxonomic status according to Andreasen 1997) have been found twice at the Late Weichselian freshwater bed at Nørre Lyngby, northern Jylland. A left mandible including M_1 was found washed out by the rain below the cliff in 1877 (Winge 1899) (Fig. 11) and a little more than a 100 years later a right P^4 was found as part of a large bone assemblage in a sand/gravel layer dated to between $11\ 590 \pm 130$ and $11\ 120 \pm 160$ ^{14}C yr BP (AAR-1509–11 and 1908–10) (Aaris-Sørensen 1995) corresponding to ca. 13.6–13.1 cal. kyr BP.

Number of dated records. – 1.

Adjacent areas. – The spread of a dry, continental steppe environment into Europe during the Weichselian was followed by an expansion of steppe elements as *S. major* (see also *D. moschata, O. pusilla, M. gregalis* and *S. tatarica*). The species reached its maximum north-western extent during the Late Glacial; additional to the Danish specimens are records from the sites of Stellmoor and Meiendorf in Holstein (Krause & Kollau 1943), Kartstein Felswand in northern Eifel (Street & Baales 1999) and from a number of cave sites in southern Britain (together with *O. pusilla*) (Yalden 1999).

Fig. 11. Left mandible of a russet ground squirrel (*Spermophilus major*) found at the Late Weichselian freshwater bed at Nørre Lyngby, N Jylland. Age: ca. 13.6–13.1 cal. kyr BP. (Photo: Geert Brovad).

Inferred time range. – ca. 14.0–11.5 cal. kyr BP.

Castor fiber

Beaver

First appearance datum. – Beaver-gnawed branches have been found in sediments dated to the Allerød period or to the transition Allerød–Younger Dryas at Nørre Lyngby, northern Jylland (Jessen & Nordmann 1915; Iversen 1942) and on the island of Sjælland at Femsølyng Mose and Favrbo Knold (Degerbøl 1928). The oldest bone remains have likewise been found in an Allerød context that is at the Final Palaeolithic site of Bromme near Sorø, Sjælland (Degerbøl 1946a).

Last appearance datum. – The latest unequivocal occurrences of *C. fiber* are recorded from the Late Bronze Age site of Hasmark Sønderby, northern Fyn (Degerbøl 1928) and from a bog near Knardrup, Måløv in northern Sjælland where a humerus has been directly radiocarbon dated to 3400 ± 45 ^{14}C yr BP (AAR-4171) corresponding to ca. 3.7–3.6 cal. kyr BP. A few Iron Age records (including the Viking Period) have been reported from Jylland and Sjælland, but it still needs to be verified whether the overall archaeological dating of these sites also include the beaver remains.

Number of dated records. – 103. Except for the sparse Allerød finds, the records are continuously dispersed between Preboreal and LAD indicating a probable Younger Dryas Induced Pause.

Inferred time range. – ca. 13.5–2.5 cal. kyr BP, most likely with a Younger Dryas Induced Pause ca. 12.6–11.4 cal. kyr BP.

Lemmus lemmus

Norway lemming

First appearance datum. – Remains have been recorded from lacustrine beds in a coastal cliff section at Kobbelgård on the island of Møn (Bennike *et al.* 1994). The deposits date to a Middle Weichselian interstadial around 30.0 cal. kyr BP (lemming bones have been directly dated to $24\ 700 \pm 5500$ and $24\ 000 \pm 5000$ ^{14}C yr BP (AAR-1012–13).

Last appearance datum. – A single molar found with other small mammal bones in a Late Weichselian freshwater bed at Nørre Lyngby, northern Jylland. The bone assemblage dates between $11\ 590 \pm 130$ and

11 120 ± 160 ^{14}C yr BP (AAR-1509–11 and 1908–10) (Aaris-Sørensen 1995) corresponding to ca. 13.6–13.1 cal. kyr BP.

Number of dated records. – 2.

Adjacent areas. – L. lemmus is recorded extensively from northern and central Europe throughout the last glacial period, from Britain and France to the Russian Plane and as far south as to Hungary and Romania (Kowalski 2001). Closest to the Danish finds are Late Glacial remains found on the site of Stellmoor in Holstein (Krause & Kollau 1943; Bratlund 1999; Benecke 2004) which proves the presence of lemming in both Bølling and Younger Dryas. The disappearance of the species marks the Pleistocene–Holocene boundary along the northern fringe of the German Mittelgebirge (Storch 1992); so, a south Scandinavian extinction is likely to have happened during the Preboreal, finally leading to its modern distribution in the Scandinavian mountains and the tundra from Lappland east to the White Sea.

Inferred time range. – Pre-LGM: Early and Middle Weichselian interstadials. Post-LGM: ca. 14.5–11.5 cal. kyr BP.

Dicrostonyx torquatus
Collared lemming

First and last appearance data. – Remains have been recorded from lacustrine beds in a coastal cliff section at Kobbelgård on the island of Møn (Bennike et al. 1994). The deposits date to a Middle Weichselian interstadial around 30.0 cal. kyr BP (lemming bones have been directly radiocarbon dated to 24 700 ± 5500 and 24 000 ± 5000 ^{14}C yr BP (AAR-1012–13).

Number of dated records. – 1.

Adjacent areas. – From the last glacial period the collared lemming has been recorded from the Russian Plane in the east to the British Isles in the west and south to the Pyrenees, Switzerland, Austria and Hungary (Yalden 1999; Kowalski 2001). It disappeared again from most of its European range at the end of the last glaciation, e.g. during the Allerød at the northern fringe of the German Mittelgebirge (Storch 1992) and during the Younger Dryas in south-east Belgium (Cordy 1991) and northern Eifel (Street & Baales 1999).

Recent evidence. – The modern distribution covers the Siberian tundra from the White Sea and eastwards

(plus the tundra in North America from Alaska to Greenland).

Inferred time range. – Pre-LGM: Early and Middle Weichselian interstadials.

Clethrionomys glareolus
Bank vole

First appearance datum. – Single specimen recovered from a submarine core south-west of Bornholm. Radiocarbon dating of terrestrial plant remains from the sample yielded an age of 8050 ± 100 ^{14}C yr BP (Ua-4859) (Heiberg & Bennike 1997) corresponding to ca. 9.1–8.7 cal. kyr BP.

Last appearance datum. – Extant.

Number of dated records. – 12, dispersed between Late Boreal and Middle Subboreal.

Adjacent areas. – Known from forest and forest-tundra habitats during the entire Late Glacial period at the northern fringe of the German Mittelgebirge (Storch 1992), in the Rhineland (Street & Baales 1999) and in the south-eastern Belgium (Cordy 1991). An expansion into southern Scandinavia probably took place in Early Preboreal.

Inferred time range. – ca. 11.4– 0 cal. kyr BP.

Arvicola terrestris
Water vole

First appearance datum. – 11 specimens found in a Late Weichselian freshwater bed at Nørre Lyngby, northern Jylland dated to between 11 590 ± 130 and 11 120 ± 160 ^{14}C yr BP (AAR-1509–11 and 1908–10) (Aaris-Sørensen 1995) corresponding to ca. 13.6–13.1 cal. kyr BP.

Last appearance datum. – Extant.

Number of dated records. – 91. Except for a Preboreal gap (see *E. europaeus*), continuously dispersed between FAD and LAD. Concerning the reliability of the finds, see *T. europaea*.

Adjacent areas. – The species is absent from northern Germany during the Early Weichselian and the Main Glaciation and first recorded here in Bølling (Storch 1992), which makes a first appearance in southern

Scandinavia probably sometimes *during* Late Bølling/ Early Allerød.

Inferred time range. – ca. 14.0–0 cal. kyr BP with a probable Younger Dryas Induced Pause ca. 12.6– 11.4 cal. kyr BP.

Microtus agrestis

Field vole

First appearance datum. – First record from a Meso- lithic site, Maglemosegårds Vænge, Vedbæk, northern Sjælland dated to Middle Atlantic (layer 5: 6120 ± 100 ^{14}C yr BP (K-3171) (Christensen 1982; Aaris-Sørensen & Andreasen 1995).

Last appearance datum. – Extant.

Number of dated records. – 18. Except for a gap in Late Subboreal (lack of Bronze Age sites with small mammal assemblages) the records are evenly dis- persed between FAD and LAD.

Adjacent areas. – Distinction between *M. agrestis* and *Microtus arvalis* is only possible if the charac- teristic 'extra' postero-internal loop on the second upper molar of *M. agrestis* is present. This makes it difficult to follow the history of the species during the last glacial period: most available fauna lists from Middle and Northern Europe lump the two species.

Recent evidence. – Common for the variety of habi- tats occupied by the field vole is a tall, dense grass cover. An Early Preboreal immigration seems likely.

Inferred time range. – ca. 11.4–0 cal. kyr BP.

Microtus arvalis

Common vole

First appearance datum. – No subfossil record.

Last appearance datum. – Extant.

Adjacent areas. – See *M. agrestis*

Recent evidence. – The northern limit of its modern European distribution includes southern and middle parts of Jylland and runs further east following the Bal- tic coast till southern Finland. The Danish islands, The British Isles and the rest of Scandinavia lie outside its range. The present distribution points at a rather late

immigration into Jylland, probably about 2500 years ago at the Subboreal/Subatlantic transition.

Inferred time range. – ca. 2.5–0 cal. kyr BP.

Microtus oeconomus

Root vole

First appearance datum. – Two specimens found in a Late Weichselian freshwater bed at Nørre Lyngby, northern Jylland dated to between 11 590 ± 130 and 11 120 ± 160 ^{14}C yr BP (AAR-1509–11 and 1908–10) (Aaris-Sørensen 1995) corresponding to ca. 13.6– 13.1 cal. kyr BP.

Last appearance datum. – Several specimens recorded from four localities on the Island of Møn and dated to Middle Holocen (Boreal-Subboreal) (Heiberg 1995).

Number of dated records. – 5.

Adjacent areas. – Distributed all over Europe during the last glacial period (see, e.g. von Koenigswald 2002).

Recent evidence. – The modern European distribu- tion is north-eastern and patchy, with isolated popula- tions close to Denmark in the Netherlands, southern Norway and central Sweden. A coherent distribution in northern Europe probably existed at least until the Subboreal/Subatlantic transition.

Inferred time range. – Pre-LGM: Early and Middle Weichselian interstadials. Post-LGM: ca. 14.5–2.5 cal. kyr BP.

Microtus gregalis

Narrow-skulled vole

First and last appearance data. – Several specimens have been recovered at a Late Weichselian freshwater bed at Nørre Lyngby, northern Jylland which is dated to between 11 590 ± 130 and 11 120 ± 160 ^{14}C yr BP (AAR-1509–11 and 1908–10) (Aaris-Sørensen 1995) corresponding to ca. 13.6–13.1 cal. kyr BP.

Number of dated records. – 1.

Adjacent areas. – The spread of a dry, continental steppe environment into Europe during the Weich- selian was followed by an expansion of steppe ele- ments such as *D. moschata*, *O. pusilla*, *S. major* and *S. tatarica*. With its modern distribution divided into two parts, a northern across the Siberian tundra from the White Sea to the far north-east and a southern along the steppe belt from the southern Urals to the Manchuria, the narrow-skulled vole partly belongs to this group. Its Weichselian distribution covered all of Europe including the Iberian Peninsula and the British Isles (Kowalski 2001) and it was still present in south-east Belgium as late as in the middle of Preboreal (Cordy 1991), in the Eifel during the Younger Dryas (Street & Baales 1999), in northern Germany during Allerød (Storch 1992) and in Belarus during the Preboreal (Motuzko & Ivanov 1996). The Danish population may very well have survived the Younger Dryas cooling and lived on into the Preboreal.

Inferred time range. – Pre-LGM: Early and Middle Weichselian interstadials. Post-LGM: ca. 14.5–11.5 cal. kyr BP.

Micromys minutus

Harvest mouse

First appearance datum. – Two specimens found in a Neolithic flint mine shaft at Hov, near Thisted in northern Jylland. A radiocarbon dating of an *A. terrestris* specimen found in close connection with the *Micromys* bones gave an age of 4220 ± 35 ^{14}C yr BP (Poz-7671) corresponding to ca. 4.8–4.7 cal. kyr BP.

Last appearance datum. – Extant.

Number of dated records. – 2 (the second recovered at a Late Viking/Early Middle Ages site at Munkerup, Bornholm (identified by Henrik Høier, unpublished ZMUC files).

Recent evidence. – The modern distribution of the species and its preference for the open cultural landscape indicates a late Holocene immigration into northern Europe. The species is still missing in Norway, Sweden (except for a small population in Värmland/Dalsland which probably originates from an accidental import) and northern Finland. Therefore, the FAD mentioned above could very well represent the first true occurrence of the species in Early Subboreal.

Inferred time range. – ca. 5.0–0 cal. kyr BP.

Apodemus agrarius

Striped field mouse

First appearance datum. – No fossil record.

Last appearance datum. – Extant.

Recent evidence. – The modern distribution covers eastern and southern Europe with the Danish range restricted to the islands of Lolland and Falster. The preferred habitats are fringes of forests, grassland, fields, gardens and marshes. It points at a late immigration into northern Germany and southern Denmark.

Inferred time range. – ca. 2.5–0 cal. kyr BP.

Apodemus flavicollis/Apodemus sylvaticus

Yellow-necked mouse/Wood mouse

First appearance datum. – First record from a Mesolithic site, Maglemosegårds Vænge, Vedbæk, northern Sjælland dated to Middle Atlantic (layer 5: 6120 ± 100 ^{14}C yr BP (K-3171) (Christensen 1982; Aaris-Sørensen & Andreasen 1995).

Last appearance datum. – Both species extant.

Number of dated records. – 23, continuously dispersed between FAD and LAD. No diagnostic osteological character exists for a safe distinction between *A. flavicollis* and *A. sylvaticus*. By measuring a large assemblage of *Apodemus* remains the presence of both species will, however, be revealed by two overlapping size groupings with *flavicollis* being larger than *sylvaticus*. The 23 Danish records all consist of single bone remains and as the size range of the two species at a given time and region is unknown the species are treated here together.

Recent evidence. – The modern Fennoscandian distribution of *A. flavicollis* covers Denmark and the southern parts of Norway, Sweden and Finland. *Apodemus sylvaticus* has the same Scandinavian distribution, but it is absent in Finland. *Apodemus flavicollis* is a forest-living species, whereas *sylvaticus* prefers the open countryside. A Preboreal immigration is most likely for both species perhaps an Early Preboreal first appearance for *sylvaticus* and a Middle Preboreal for *flavicollis*.

Inferred time range. – ca. 11.0–0 cal. kyr BP (combined).

Rattus rattus

Black rat

First appearance datum. – Three parts of a skull found during excavations of a medieval house in

Kompagnistræde in Næstved (AD 1250–1400) (identified by K. Rosenlund, unpublished ZMUC files) and eight specimens (including skull fragments and mandibles) found during the excavation of the medieval castle Lykkesholm, Langeland (1260–1440) (Gelskov 2005).

Last appearance datum. – Extant, although only as sporadic occurrences of imported specimens in a few ports (2000–2003: only in the town of Nakskov and Nykøbing Falster, Lodal 2007).

Number of dated records. – 6, ranging from AD 1250 to late 1600s. *Rattus rattus* is presumed to be native to SE Asia. According to historical records it was the only rat present in Denmark throughout the medieval period and until the beginning of the 1700s (Lodal 2007) when it was gradually replaced by *Rattus norvegicus*. Therefore, ten records of *Rattus* sp. known from Jylland, Fyn, Sjælland and Bornholm and dated to ca. AD 1000–1600 should most likely be added to the six identified *R. rattus* specimens mentioned above.

Adjacent areas. – The true first appearance of the black rat should nevertheless be expected much earlier, probably during Roman time (ca. AD 0–400) as in Britain (Yalden 1999, pp. 124–125) as a result of increasing trade activity between the Middle East, the Mediterranean and northern Europe.

Inferred time range. – ca. 2.0–0 cal. kyr BP.

Rattus norvegicus

Brown rat

First appearance datum. – Only two positive identifications, a mandible and a skull both with molars, found during excavations in the city of Copenhagen (Vor Frelsers Kirke, AD 1650–1889, identified by K. Rosenlund, unpublished ZMUC file and Esplanaden, AD 1690–1750, I. B. Enghoff, unpublished manuscript).

Last appearance datum. – Extant.

Number of dated records. – 2.

Adjacent areas. – According to historical records the brown rat seems to have spread from its homeland in central Asia into Europe as late as in the early 1700s (Winge 1908 (about the earliest Danish records); Yalden (1999, p. 183: about western Europe in general). Heinrich (1976), however, reports earlier finds of *R. norvegicus* in Schleswig–Holstein from AD 800–900.

Inferred time range. – ca. AD 1700 to present time.

Mus musculus

House mouse

First appearance datum. – Earliest remains come from two Pre-Roman Iron Age settlements in northern Jylland (ca. 2.4–1.8 cal. kyr BP): Smedegård north of Thisted in northern Jylland (Raahauge 2002) and Mellemholm near Ålborg (identified by K. Rosenlund, unpublished ZMUC files).

Last appearance datum. – Extant.

Number of dated records. – 7.

Adjacent areas. – Like the black rat the house mouse is also an introduction from Asia. According to Brothwell (1981) the earliest European records are from Bronze Age sites in the Mediterranean and Yalden (1999) shows that it was certainly present in Britain by the Iron Age. This is in line with the modest Danish records.

Inferred time range. – ca. 2.0–0 cal. kyr BP.

Muscardinus avellanarius

Common dormouse

First appearance datum. – No fossil record.

Last appearance datum. – Extant.

Recent evidence. – The common dormouse lives in deciduous and mixed forests especially the forest edge and understorey shrubs (Mitchell-Jones *et al.* 1999). Southern Sweden is the northernmost limit of its modern range and the Danish distribution is patchy and restricted to certain areas in central, southern and south-western Sjælland, southern Fyn and south-eastern Jylland (Vilhelmsen 2007). A coherent distribution in central and northern Europe is likely to have been established during Early–Middle Preboreal.

Inferred time range. – ca. 11.0–0 cal. kyr BP.

Sicista cf. betulina

Northern birch mouse

First appearance datum. – A right mandible, recovered at a Late Weichselian freshwater bed at Nørre

Lyngby, northern Jylland and dated to between 11 590 ± 130 and 11 120 ± 160 ^{14}C yr BP (AAR-1509–11 and 1908–10) (Aaris-Sørensen 1995) corresponding to ca. 13.6–13.1 cal. kyr BP.

Last appearance datum. – Extant.

Number of dated records. – 2 (the second is a fragment of a mandible found on a Pre-Roman Iron Age settlement at Smedegård north of Thisted in northern Jylland dated to ca. 2.4–1.8 cal. kyr BP, Raahauge 2002).

Recent evidence. – Today the birch mouse occupies a variety of habitats in the forest, steppe–forest and mountain forest zones of Europe and Asia (Mitchell-Jones *et al.* 1999). The present Danish distribution is part of the western border of the European range and covers two areas, one in NW Jylland and another in a belt across S Jylland. A likewise patchy distribution is seen in Norway, Sweden, Finland, northernmost Germany and between the Alps and the Carpathians and documents the relict character of the recent distribution (Mitchell-Jones *et al.* 1999; Jensen & Møller 2007).

Inferred time range. – ca. 13.5–0 cal. kyr BP with a probable Younger Dryas Induced Pause ca. 12.6–11.4 cal. kyr BP.

Carnivora

Canis lupus

Wolf

First appearance datum. – A mandible found in a clay pit at Allerød Tileworks, N Sjælland. The mandible has been radiocarbon dated to 10 530 ± 75 ^{14}C yr BP (AAR-4172) corresponding to ca. 12.7–12.4 cal. kyr BP (Fig. 12).

Last appearance datum. – Traditionally, the last Danish wolf is claimed to have been shot the night of June 20 in 1813 by the gamekeeper at Estvadgård near Skive in northern Jylland. Most likely, however, this and other late wolves are stray animals from northern Germany. According to historical records the last native wolf populations disappear from the islands at the end of the 1500s and in Jylland at the end of the 1600s (Erslev 1871).

Number of dated records. – 46, continuously dispersed between FAD and LAD except for a gap in Preboreal.

Fig. 12. Mandible of a wolf (*Canis lupus*) found in a clay pit at Allerød, N Sjælland. Age: ca. 12.7–12.4 cal. kyr BP. (Photo: Geert Brovad).

The first appearance of the wolf in the deglaciated southern Scandinavia undoubtedly coincided with the immigration of its principal prey, the reindeer around 15.0–14.5 cal. kyr BP. The absence of evidence from the Preboreal should once again be ascribed to the lack of faunal remains from the archaeological sites from this period.

Adjacent areas. – The wolf was present all over Europe during the last glacial period (see, e.g. Kahlke 1999; von Koenigswald 2002). The lack of evidence in Preboreal Denmark is compensated for by wolf records from Late Younger Dryas Stellmoor in Holstein (Krause & Kollau 1943; Benecke 2004), from Early Preboreal at Thatcham in Berkshire (King 1962), from Middle Preboreal at Star Carr in Yorkshire (Fraser & King 1954) and from Late Preboreal Friesack in Brandenburg (Gramsch 2000).

Inferred time range. – Pre-LGM: Early and Middle Weichselian interstadials. Post-LGM: ca. 14.5 cal. kyr BP to AD 16–1700.

Vulpes lagopus (Alopex lagopus)

Arctic fox

Remarks. – No fossil record in Denmark. A left mandible has, however, been found in a kettle hole at Hässleberga in south-western Skåne together with numerous reindeer bones in a calcareous detritus gyttja of Late Glacial origin (Larsson *et al.* 2002) (Fig. 13). Because of the palaeogeographical setting (Fig. 2H) this Swedish find is as representative of the Danish as it is of the Swedish Late Glacial fauna. The arctic fox is also known from all over Europe during the last glacial period (see, e.g. Kahlke 1999; von Koenigswald 2002).

Fig. 13. Left mandible of an arctic fox (*Vulpes lagopus*) found in a kettle hole at Hässleberga in SW Skåne. Age: ca. 14.5–11.0 cal. kyr BP. Scale bar = 5 cm. (Photo: Geert Brovad).

Inferred time range. – Pre-LGM: Early and Middle Weichselian interstadials. Post-LGM: ca. 14.5–11.0 cal. kyr BP.

Vulpes vulpes

Red fox

First appearance datum. – Mesolithic (Maglemosian) sites of Mullerup, Sværdborg and Holmegård, ca. 9.5 cal. kyr BP (Winge 1903, 1919, 1924; Aaris-Sørensen 1976).

Last appearance datum. – Extant.

Number of dated records. – 118, continuously dispersed between FAD and LAD. For reasons discussed in the case of *E. europaeus* the lack of Preboreal bone remains is not reliable evidence of absence of the species during that period.

Adjacent areas. – Preboreal records are known from nearby localities in northern Germany. At the Upper Palaeolithic site of Stellmoor in Holstein Bratlund (1999) has recognized that the Ahrensburgian assemblage has an admixture of Post Glacial elements including bones of *V. vulpes* which have been directly radiocarbon dated to 9680 ± 90 ^{14}C yr BP (OxA-2875). Calibrated, this corresponds to ca. 11.2–10.8 cal. kyr BP. Further, contemporary remains of the red fox have also been found at the site of Friesack in Brandenburg (Zeitstube I) (Gramsch 2000).

Inferred time range. – ca. 11.4–0 cal. kyr BP.

Ursus arctos

Brown bear

First appearance datum. – The two earliest finds come from Late Glacial Allerød layers, at Faurbo

Knold in north-western Sjælland and at the coastal cliff at Nørre Lyngby, northern Jylland. The latter has been indirectly radiocarbon dated to 11 430 ± 160 ^{14}C yr BP (K-2900) (Aaris-Sørensen & Petersen 1984; Bondesen & Lykke-Andersen 1978) corresponding to ca. 13.4 cal. kyr BP.

Last appearance datum. – Remains from the Middle Neolithic (Pitted Ware) site of Kainsbakke west of Grenå in eastern Jylland (Richter 1991). The *U. arctos* remains have been indirectly radiocarbon dated (aurochs) to 4060 ± 50 ^{14}C yr BP (Ua-24709) corresponding to ca. 4.8–4.4 cal. kyr BP.

Number of dated records. – 41. Records are continuously dispersed between Preboreal and LAD, except for only two Allerød finds which probably indicate a Younger Dryas Induced Pause. *Regional* differences can be seen after the Early Atlantic transgressions around 8.0 cal. kyr BP (see the section on Islands).

From the Iron Age (including the Viking Period) (ca. 2.5–1.0 cal. kyr BP) *U. arctos* remains have been found all over the country in urn graves, war-booty sacrifices and trading posts. Except for a single perforated tooth pendant and a single metacarpal all other remains are distal phalanges supposed to derive from bear skins imported from Norway, Sweden or eastern Europe (Møhl 1971b, 1977). Therefore, none of these younger finds are believed to represent a Danish bear population.

Inferred time range. – ca. 13.5–4.6 cal. kyr BP, most likely with a Younger Dryas Induced Pause ca. 12.6–11.4 cal. kyr BP (for regional differences see the section on Islands).

Ursus maritimus

Polar bear

First and last appearance data. – Only one single record, a left mandible found during exploitation of gravel in the bank of Kjul Å at Asdal near Hirtshals in northern Jylland (Nordmann & Degerbøl 1930; Aaris-Sørensen & Petersen 1984) (Fig. 14). The specimen has been radiocarbon dated to 11 100 ± 160 ^{14}C yr BP (K-3741); calibrated and corrected for marine reservoir effect (−400 years) this corresponds to ca. 12.9–12.4 cal. kyr BP.

Number of dated records. – 1.

Adjacent areas. – Remains of additional seven polar bears have been found in Scandinavia with six on the Swedish west coast (Berglund *et al.* 1992) and a nearly complete skeleton at Finnøy north of Stavanger in

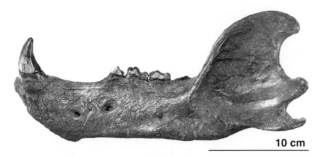

Fig. 14. Left mandible of a polar bear (*Ursus maritimus*) found at Kjul Å near Asdal in northern Jylland. Age: ca. 12.9–12.4 cal. kyr BP. (Photo: Geert Brovad).

south-western Norway (Blystad *et al.* 1983). Like the Danish these are all of Late Glacial origin ranging from ca. 15.4–12.1 cal. kyr BP. Furthermore, polar bears have been reported from interstadial faunas further north along the Norwegian west coast: Hamnsundhelleren north of Ålesund, age ca. 36.0–28.0 ^{14}C kyr BP (Hufthammer 2001) and Nordcemgrotta, Kjøpsvik, age ca. 22.0 and >70.0 ^{14}C kyr BP (Lauritzen *et al.* 1996).

Inferred time range. – Pre-LGM: whenever a marine palaeo-Kattegat–Skagerrak existed. Post-LGM: ca. 18.0–11.7 cal. kyr BP.

Mustela erminea

Stoat

First appearance datum. – A skull fragment and three mandibles have been found on a Pre-Roman Iron Age settlement at Smedegård north of Thisted in northern Jylland. A direct radiocarbon date of one of the mandibles gave a calibrated age of ca. 2.0 cal. kyr BP (KIA-8831) (Raahauge 2002).

Last appearance datum. – Extant.

Number of dated records. – 1.

Adjacent areas. – Subfossil remains of this small mustelid are extremely rare also outside Denmark. It has, however, been found all over Europe during the last glacial period (see, e.g. von Koenigswald 2002) and a Scandinavian find comes from a core sample from the island of Andøya in northern Norway in a stratum dated to ca. 15.0 ^{14}C kyr BP (Fjellberg 1978). Other Late Glacial records are known from several European countries including Ireland, Belgium and France (Sommer & Benecke 2004). Post Glacial specimens are known from the Mesolithic site of Friesack in Brandenburg dated to ca. 10.3–9.6 cal. kyr BP

(Zeitstufe III) (Gramsch 2000) and at the Ahrensburgian site of Kartstein Felswand in northern Eifel dated to the cold Younger Dryas stadial (Street & Baales 1999).

Recent evidence. – The modern distribution in Europe covers the northern and central parts from the tundra to the deciduous forest zones.

Inferred time range. – Pre-LGM: Early and Middle Weichselian interstadials. Post-LGM: ca. 14.0–0 cal. kyr BP.

Mustela nivalis

Weasel

First appearance datum. – Mandibles, scapula and humerus of a single individual found in a small bronze box in a female grave near Frederikssund in northern Sjælland. The grave dates to Early Bronze Age, ca. 2.5 cal. kyr BP (Winge 1904). Only other subfossil records are a mandible found on an Early Viking Age site at Snubbekorsgård, Smørum west of Copenhagen (ca. AD 800–900) (unpublished report by A. B. Gotfredsen in ZMUC files) and a left mandible found on a Late Medieval site at Stakhaven, Dragør (ca. AD 1300–1350) (Rosenlund 1979).

Last appearance datum. – Extant.

Number of dated records. – 3.

Adjacent areas. – Subfossil remains are extremely rare also outside Denmark. It has, however, been found all over Europe during the last glacial period (see e.g. von Koenigswald 2002). Late Glacial records are known from at least 29 sites in Europe including archaeological sites in Germany and France (Sommer & Benecke 2004). Post Glacial records are known from the Mesolithic site of Friesack in Brandenburg dated to ca. 10.3–9.6 cal. kyr BP (Zeitstufe III) (Gramsch 2000), from the Ahrensburgian site of Kartstein Felswand in northern Eifel dated to the cold Younger Dryas interstadial (Street & Baales 1999) and from the Palaeolithic site of Andernach in the Neuwied Basin dated to ca. 15.5 cal. kyr BP (Street 1997).

Recent evidence. – The modern distribution in Europe covers the entire mainland and the British Isles except Ireland.

Inferred time range. – Pre-LGM: Early and Middle Weichselian interstadials. Post-LGM: ca. 14.0–0 cal. kyr BP.

Mustela putorius

Western polecat

First appearance datum. – Mesolithic (Maglemosian) sites of Lundby, Mullerup, Sværdborg and Holme-gård, ca. 9.5 cal. kyr BP (Winge 1903, 1919, 1924; Aaris-Sørensen 1976; Rosenlund 1980).

Last appearance datum. – Extant.

Number of dated records. – 20. The relatively few records are continuously dispersed between FAD and LAD except for a lack of finds dated to Late Subboreal. Concerning the lack of Preboreal evidence see *E. euro-paeus. Regional* differences can be seen after the Early Atlantic transgressions around 8.0 cal. kyr BP (see the section on Islands).

Adjacent areas. – The only find from adjacent areas comes from the Mesolithic site of Friesack in Bran-denburg dated to ca. 10.3–9.6 cal. kyr BP (Zeitstufe III) (Gramsch 2000).

Inferred time range. – ca. 11.4–0 cal. kyr BP (for regio-nal differences see the section on Islands).

Martes foina

Beech marten

First appearance datum. – No subfossil record.

Last appearance datum. – Extant.

Adjacent areas and recent evidence. – The modern distribution covers the whole of mainland Europe (besides central and western Asia) including Den-mark and southern Estonia as the northernmost limit. The absence of the beech marten in England, Sweden, Finland and Norway points at a late Holo-cene spread into northern Europe. This is supported by the very few subfossil records known from Eur-ope (Sommer & Benecke 2004) which show the first occurrences in central Europe, especially France and Italy, as late as in the Atlantic. It seems obvious that the expansion of the beech marten follows the spread of the Neolithic culture and its opening of the landscape. Unfortunately, the records are too few to test this assumption. In fact the first secure records from the northern European lowlands date to the Middle Ages (Requate 1956; Sommer & Bene-cke 2004). In Denmark it is supposed to be a very late immigrant arriving some times during the last 2000 years.

Inferred time range. – ca. 2.0–0 cal. kyr BP.

Martes martes

Pine marten

First appearance datum. – Mesolithic (Maglemosian) sites of Lundby, Mullerup, Sværdborg and Holme-gård, ca. 9.5 cal. kyr BP (Winge 1903, 1919, 1924; Aaris-Sørensen 1976; Rosenlund 1980).

Last appearance datum. – Extant.

Number of dated records. – 91, continuously dis-persed between FAD and LAD. Concerning the lack of Preboreal evidence, see *E. europaeus.*

Adjacent areas. – Remains are known from the Mesolithic site of Friesack in Brandenburg dated to ca. 11.0–10.7 cal. kyr BP (Zeitstufe I) (Gramsch 2000), Star Carr in Yorkshire ca. 11.0–10.6 cal. kyr BP (Fraser & King 1954) and Thatcham in Berkshire ca. 11.5–11.0 cal. kyr BP (King 1962).

Inferred time range. – ca. 11.4–0 cal. kyr BP.

Gulo gulo

Wolverine

First and last appearance data. – Only one single record, a skull fragment found on the Palaeolithic site of Bromme near Sorø, Sjælland (Degerbøl 1946a) (Fig. 15). Indirectly (*A. alces*) radiocarbon dated to

Fig. 15. Part of the upper jaw and the snout of a wolverine (*Gulo gulo*) found at a Palaeolithic settlement at Bromme near Sorø, Sjæl-land. Age: ca. 12.8 cal. kyr BP. (Photo: Geert Brovad).

10 720 ± 90 ^{14}C yr BP (AAR-4539) corresponding to ca. 12.8–12.7 cal. kyr BP.

Adjacent areas. – The species is likewise reported from the Upper Palaeolithic site of Meiendorf in Holstein (Krause 1937) dated to ca. 14.5–14.0 cal. kyr BP (Fischer & Tauber 1987; Grimm & Weber 2008) and from a number of caves and rock shelters in southern England belonging to both the Full (Devensian) Glacial and the Late Glacial (Yalden 1999). On the European mainland it is known throughout the full glacial period of the Weichselian as far south as to the Pyrenees (von Koenigswald 2002).

Inferred time range. – Pre-LGM: Early and Middle Weichselian interstadials. Post-LGM: ca. 14.5–11.0 cal. kyr BP.

Meles meles

Badger

First appearance datum. – Limb bones found in a marl pit, Overgårds Mergelleje north-east of Vejle in eastern Jylland, directly radiocarbon dated to 9965 ± 60 ^{14}C yr BP corresponding to ca. 11.6–11.3 cal. kyr BP.

Last appearance datum. – Extant.

Number of dated records. – 51, continuously dispersed between FAD and LAD. *Regional* differences can be seen after the Early Atlantic transgressions around 8.0 cal. kyr BP (see the section on Islands).

Adjacent areas. – Other Early Preboreal occurrences of badger are recorded from the Mesolithic sites of Thatcham in Berkshire (King 1962) and Star Carr in Yorkshire (Fraser & King 1954) and it is known as early as in the Allerød Interstadial in the Central Rhineland Neuwied Bassin (site of Niederbieber) (Baales & Street 1996; Street 1997).

Inferred time range. – ca. 11.4–0 cal. kyr BP (for regional differences see the section on Islands).

Lutra lutra

Otter

First appearance datum. – Mesolithic (Maglemosian) sites of Lundby, Sværdborg and Holmegård, ca. 9.5 cal. kyr BP (Winge 1919, 1924; Aaris-Sørensen 1976; Rosenlund 1980).

Last appearance datum. – Extant.

Number of dated records. – 101, continuously dispersed between FAD and LAD. Concerning the lack of Preboreal evidence, see *E. europaeus.*

Adjacent areas. – First appearance data from adjacent areas come from the Mesolithic sites of Friesack in Brandenburg (Zeitstube II), ca. 10.6–10.3 cal. kyr BP (Gramsch 2000) and Hohen Viecheln in NW Mecklenburg dated to the transition Preboreal/Boreal ca. 10.3 cal. kyr BP (Gehl 1961).

Inferred time range. – ca. 11.4–0 cal. kyr BP.

Felis silvestris

Wild cat

First appearance datum. – Mesolithic (Maglemosian) sites of Lundby, Mullerup, Sværdborg and Holmegård, ca. 9.5 cal. kyr BP (Winge 1903, 1919, 1924; Aaris-Sørensen 1976; Rosenlund 1980).

Last appearance datum. – Remains from the site of Næsbyholm south of Sorø, Sjælland, dated to early Roman Iron Age ca. 2.0 cal. kyr BP (U. Møhl, unpublished ZMUC files).

Number of dated records. – 82, continuously dispersed between FAD and LAD. Concerning the lack of Preboreal evidence, see *E. europaeus.*

Adjacent areas. – First appearance data from adjacent areas come from the Mesolithic site of Friesack in Brandenburg (Zeitstube I), ca. 11.0–10.7 cal. kyr BP (Gramsch 2000) and in Britain from two early Mesolithic sites in Berkshire, Thatcham, ca. 11.5–11.0 cal. kyr BP (King 1962) and Faraday Road, ca. 10.7–10.6 cal. kyr BP (Ellis *et al.* 2003).

Inferred time range. – ca. 10.8–2.0 cal. kyr BP.

Lynx lynx

Lynx

First appearance datum. – Two specimens found on the Mesolithic site of Øgaarde south-east of Holbæk, Sjælland, pollen dated to Late Boreal, ca. 9.5 cal. kyr BP (Degerbøl 1943; Troels-Smith 1943).

Last appearance datum. – Remains found in the Late Mesolithic Ertebølle sequence on the site of Bjørnsholm south of Løgstør in northern Jylland, ca. 7.0–6.0 cal. kyr BP (Bratlund 1993) and on the Late Mesolithic/Early Neolithic site of Dyrholmen

Fig. 16. Skull of a lynx (*Lynx lynx*) found at Slagsmose near Næstved, Sjælland. Age: ca. 9.0–8.0 cal. kyr BP. Scale bar = 5 cm. (Photo: Geert Brovad).

south-east of Randers in eastern Jylland, ca. 7.0–5.5 cal. kyr BP (Degerbøl 1942).

Number of dated records. – 16, the vast majority belonging to the Middle and Late Atlantic (Fig. 16). Concerning the lack of Preboreal evidence see *E. europaeus*. Despite the few records regional differences seem nevertheless recognizable after the Early Atlantic transgressions around 8.0 cal. kyr BP (see the section on Islands).

Adjacent areas. – Late Glacial and Early Holocene remains have been found in Britain in Gough's Cave, Somerset and in Dog Hole Fissure, Derbyshire, directly radiocarbon dated to 12 650 ± 120 and 9570 ± 60 ^{14}C yr BP respectively (Hedges *et al.* 1994; Bronk Ramsey *et al.* 2002). The first Holocene records in northern Germany is known from the Mesolithic site of Hohen Viecheln in NW Meclenburg dated to the Preboreal/Boreal transition ca. 10.3 cal. kyr BP (Gehl 1961).

Inferred time range. – ca. 10.8–5.0 cal. kyr BP.

Pinnipedia

Odobenus rosmarus

Walrus

First appearance datum. – Only five records in all. Four specimens (three dredged from the Danish part of the North Sea and one excavated in a gravel pit in northern Jylland) have been directly radiocarbon dated to late Middle Weichselian and early Late Weichselian. The oldest date of 30 880 + 1270/−1110 ^{14}C yr BP (K-3726) (Møhl 1985) belongs to a specimen found 40 nautical miles NW of Lyngvig Fyr and the youngest of 23 550 ± 460 ^{14}C yr BP (K-4473) (Aaris-Sørensen 1998) belongs to a specimen found in a gravel pit in northern Jylland near Kirkholt, Hjørring.

Last appearance datum. – In the present time vagrants are occasionally observed on the Atlantic coasts south to the Gulf of Biscay (Mitchell-Jones *et al.* 1999) including eight records from the Danish North Sea coast in the period 1926–2005 (Tougaard 2007). Rare visits like these have probably happened throughout the Holocene.

Number of dated records. – 4.

Inferred time range. – Pre-LGM: Whenever a marine palaeo-Kattegat–Skagerrak existed. Post-LGM: ca. 18.0–11.7 (probably frequent) and ca. 11.7–0 cal. kyr BP (a rare visitor).

Phoca groenlandica

Harp seal

First appearance datum. – A series of 24 selected *P. groenlandica* specimens have been directly radiocarbon dated (Bennike *et al.* 2008) giving an oldest date of 5690 ± 50 ^{14}C yr BP (LuS-6129). Corrected for marine reservoir effect (−400 years) and then calibrated, this date corresponds to ca. 6.2–5.9 cal. kyr BP.

Last appearance datum. – Extant, very rare, occasional invasions from the north-eastern Atlantic stock.

Number of dated records. – 47 falling in three age groups, one centred around 6.0 cal. kyr BP, another around 4.7 cal. kyr BP and the third scattered between ca. 3.5 cal. kyr BP and the present time (Bennike *et al.* 2008).

Adjacent areas. – A number of Late Glacial and Early Holocene harp seal remains have been found on the Swedish west coast and their radiocarbon dates range between 11 470 ± 175 and 9885 ± 155 ^{14}C BP (St-4403 and 4811) (Fredén 1984). Calibrated and corrected for marine reservoir effect this corresponds to ca. 13.3–11.4 cal. kyr BP. The harp seal reappears on the Swedish west coast and in the Baltic Sea in Early Subboreal (Lepiksaar 1964; Fredén 1975, 1984;

Ukkonen 2002; Storå & Ericson 2004) which is in agreement with the oldest range of the Danish record.

Inferred time range. – Pre-LGM: Whenever a marine palaeo-Kattegat–Skagerrak existed. Post-LGM: ca. 18.0–11.7 and ca. 6.0–0 cal. kyr BP.

Phoca hispida

Ringed seal

First appearance datum. – Two specimens are directly radiocarbon dated to the time *before* LGM with ages around 42.0 ^{14}C kyr BP. A humerus was found in a gravel pit at Egegaard near Kgs. Lyngby north of Copenhagen (42 300 ± 2000 ^{14}C yr BP, LuS-7371) and a metacarpale was found during a well digging at Ammendrup south of Helsinge in northern Sjælland (42 200 ± 1000 ^{14}C yr BP, LuS-7691). First occurrences *after* LGM are known by an almost complete skeleton found at Hørmested between Hjørring and Frederikshavn in northern Jylland which has been directly radiocarbon dated to 14 410 ± 80 ^{14}C yr BP (LuS-7687) corresponding to ca. 17.1–16.7 cal. kyr BP (Fig. 17). Of similar age is a thoracic vertebra found in a clay pit at Nivå, north-eastern Sjælland and radiocarbon dated to 14 110 ± 250 ^{14}C yr BP (Ua-1023) (Lagerlund & Houmark-Nielsen 1993).

Last appearance datum. – Extant, very rare, occasionally individuals will stray into Danish waters from both the North Atlantic and the Baltic populations.

Number of dated records. – 39 (Pre-LGM: 6. Post-LGM: 33).

Inferred time range. – Pre-LGM: Whenever a marine palaeo-Kattegat–Skagerrak existed. Post-LGM: ca. 18.0–11.7 (probably frequent) and ca. 11.7–0 cal. kyr BP (a rare visitor).

Phoca vitulina

Common seal

First appearance datum. – Remains found on two Mesolithic sites on the coast of Øresund between Copenhagen and Helsingør (Bloksbjerg & Nivå) dated to Kongemose/Ertebølle culture around 8.0–6.0 cal. kyr BP.

Last appearance datum. – Extant.

Number of dated records. – 19. Except for a gap in Late Subboreal, the 19 dates are evenly dispersed between FAD and LAD.

Adjacent areas. – Subfossil finds from adjacent areas are extremely scarce with the oldest being found on Middle Neolithic (Pitted Ware) sites on the Swedish west coast (Lepiksaar 1964). This corresponds to the Early Subboreal around 5.0 cal. kyr BP and the species is generally believed to be a rather late immigrant into the northern European coastal waters (Sommer & Benecke 2003).

Inferred time range. – ca. 8.0–0 cal. kyr BP.

Halichoerus grypus

Grey seal

First appearance datum. – Remains found on Mesolithic sites belonging to the Kongemose/Ertebølle culture around 8.0–6.0 cal. kyr BP: Øster Jølby on the island of Mors in Limfjorden, northern Jylland, Carstensminde on the island of Amager, Bloksbjerg and Vedbæk Boldbaner on the coast of Øresund north of Copenhagen, Villingebæk Ø on the north coast of Sjælland and finally Argusgrunden in Guldborgsund between the islands of Lolland and Falster.

Last appearance datum. – Extant.

Number of dated records. – 98, continuously dispersed between FAD and LAD.

Adjacent areas. – Direct radiocarbon dates of two grey seals from the Mesolithic site of Tågerup on the

Fig. 17. Almost complete skeleton of a ringed seal (*Phoca hispida*) found at Hørmested between Hjørring and Frederikshavn, N Jylland. Age: 17.1–16.7 cal. kyr BP. (Photo: Geert Brovad).

west coast of Skåne fall within the Danish FAD values. They are dated to 8095 ± 90 and 7405 ± 85 ^{14}C yr BP (Ua-25206 and -25204); after being corrected for marine reservoir effect (−400 years) and calibrated they range between ca. 8.6 and 7.7 cal. kyr BP (Eriksson & Magnell 2001). A much earlier occurrence in the Kattegat area is, however, documented by three Early Holocene specimens found on the coast of Bohuslän. These are also directly radiocarbon dated with an oldest date of 9920 ± 215 ^{14}C yr BP, which, corrected and calibrated, corresponds to the interval between 11.9 and 11.1 cal. kyr BP (Fredén 1984).

Inferred time range. – ca. 11.7–0 cal. kyr BP.

Proboscidea

Mammuthus primigenius

Woolly mammoth

First appearance datum. – Tusk, molar and pelvis fragments with non-finite dates (>ca. 40.0 ^{14}C kyr BP) (K-4587, K-4191 and K-4188) found at Rosmos north of Ebeltoft in eastern Jylland and Sønder Kollemorten and Sønder Omme both between Vejle and Skjern in Middle Jylland (Aaris-Sørensen *et al.* 1990; Aaris-Sørensen 2006). The specimens probably represent interstadial faunas from the Early and early Middle Weichselian. A molar found at Rolsøgård south of Rønde in eastern Jylland has the oldest finite date of 41 500 ± 700 ^{14}C yr BP (OxA-1090) (Aaris-Sørensen 2006).

Last appearance datum. – A molar found at Højballegård north of Horsens in eastern Jylland dated to 19 940 ± 120 ^{14}C yr BP (LuS-7415) (Aarppe & Karhu in press).

Number of dated records. – 20. Besides the three non-finite and the FAD the others are evenly distributed between ca. 35.0 and 20.0 ^{14}C kyr BP (Fig. 18).

Adjacent areas. – A northward expansion of the European mammoth population into southern Scandinavia after the LGM is documented by several dates of a tusk found at Lockarp, Skåne. Berglund *et al.* (1976) report a dating of 13 360 ± 95 ^{14}C yr BP (Lu-796) of the Lockarp specimen and Kjær *et al.* (2006) has recently obtained a date of 13 310 ^{14}C yr BP (Poz-3941). Besides, four young Late Glacial dates between ca. 14.0 and 13.0 ^{14}C kyr BP are known from northern Poland (Kubiak 1980; Wojtal 2007) and in Britain an adult and three young mammoths were discovered in a kettle hole at Condover in Shropshire and dated to ca. 12.3 ^{14}C kyr BP (Coope & Lister 1987) just as two specimens from caves of the Creswell Crags in northern Midlands and one from Gough's Cave in Somerset yielded dates between ca. 12.4 and 12.2 ^{14}C kyr BP (Currant *et al.* 1989; Housley 1991; Lister 1991). So, the last occurrence of mammoth in Britain, northern Poland and southern Scandinavia range between ca. 17.0 and 14.0 cal. kyr BP.

Inferred time range. – Pre-LGM: Early and Middle Weichselian interstadials. Post-LGM: ca. 17.0–14.0 cal. kyr BP.

Perissodactyla

Equus ferus

Wild horse

First appearance datum. – A distal part of a tibia found in a gravel pit at Vejrhøj near Vojens, SE Jylland – directly radiocarbon dated to 38 990 + 1100/−970 (KiA-19280) (Aaris-Sørensen 2003).

First appearance *after* LGM is documented by two specimens from Ulvmose south-west of Frederikshavn and Over Næsgårds Mose at Brovst in northern Jylland dated to 10 010 ± 170 (K-5748) and 9980 ± 75 ^{14}C yr BP (AAR-4543), corresponding to ca. 11.7–11.3 cal. kyr BP.

Fig. 18. Two of the best preserved Danish mammoth (*Mammuthus primigenius*) remains: central part of a tusk found at Hedehusene, E of Copenhagen, 126 cm long and with a diameter of 17 cm, and a complete molar from Kelstrup Strand S of Haderslev. Age: undated, finite dates of other specimens range between ca. 35.0 and 20.0 ^{14}C kyr BP. (Photo: Geert Brovad).

Last appearance datum. – Specimen found on the Middle Neolithic site of Lindskov near Horsens, dated to 4520 ± 65 ^{14}C yr BP (K-2652) (Davidsen 1978) corresponding to ca. 5.3–5.0 cal. kyr BP.

Number of dated records. – 16. Nine of the 15 records *after* LGM are Preboreal and found in non-archaeological contexts, whereas the rest have been found on Late Mesolithic/Early–Middle Neolithic sites belonging to Late Atlantic/Early Subboreal. As the earliest undisputed evidence of domestic horses are chariot burials dating to ca. 4.0 cal. kyr BP from Krivoe Ozero on the Ural Steppe (Jansen *et al.* 2002) these Late Mesolithic/Early–Middle Neolithic Danish horses are interpreted as true wild horses.

Adjacent areas. – Horses are well documented from Weichselian deposits in the whole of central and northern Europe (e.g. Kahlke 1999; von Koenigswald 2002). Besides, remains have been found in Late Glacial kettle holes at Hässleberga in south-western Skåne where six dates range between ca. 13.2 and 12.2 cal. kyr BP (Larsson *et al.* 2002).

Inferred time range. – Pre-LGM: Early and Middle Weichselian interstadials. Post-LGM: ca. 13.5–10.4 with a probable Younger Dryas Induced Pause ca. 12.6–11.4. After a long period of absence there is a new and last appearance of the wild horse between ca. 6.0 and 5.0 cal. kyr BP.

Coelodonta antiquitatis

Woolly rhinoceros

Remarks. – Remains of rhinoceros are extremely rare in Denmark. Only five specimens, all discovered in glacial sediments, are available and only two of them can be identified to species. A molar found at Søby south of Ikast in Middle Jylland has the characteristic rugose enamel of *C. antiquitatis* and a fragment of a left mandible found in a gravel pit at Seest near Kolding, SE Jylland, can be assigned with certainty to Merck's rhinoceros, *Dicerorhinus kirchbergensis* and is believed to date to the Eemian. The rest are fragments of limb bones which cannot be morphologically identified to species and as Eemian bone remains can be found redeposited in glacial sediments together with Weichselian elements, neither can they be startigraphically/ecologically identified. An identification by means of ancient DNA analyses have been attempted without success (E. Willerslev, personal communication 2009) and radiocarbon datings only gave a non-finite date of >44 000 ^{14}C yr BP (Seest, SE Jylland, LuS-7375) and an apparent age

of 41 500 ± 1800 ^{14}C yr BP (Seest, SE Jylland, LuS-7377).

Adjacent areas. – Widely distributed on the Weichselian mammoth steppe in Europe and represented in especially large numbers during the LGM in Middle Europe (e.g. Kahlke 1999; von Koenigswald 2002).

Inferred time range. – Early and Middle Weichselian interstadials.

Artiodactyla

Sus scrofa

Wild boar

First appearance datum. – Specimen found in Årsballe Mose, Bornholm and radiocarbon dated to 9120 ± 120 ^{14}C yr BP (K-4637) corresponding to ca. 10.5–10.2 cal. kyr BP.

Last appearance datum. – Locally extinct during the AD 1700s. The last wild boar is claimed to have been shot in 1801 near Silkeborg, Jylland (Weismann 1931).

Number of dated records. – 222, continuously dispersed between FAD and LAD.

Adjacent areas. – Early to Middle Preboreal wild boar remains are known from Mesolithic sites in northern Germany and Britain. The earliest occurrence is documented by a direct radiocarbon date of a wild pig at the site of Potsdam-Schlaatz, Brandenburg, giving an age of 9956 ± 54 ^{14}C yr BP (KIA-9563) (Benecke *et al.* 2002), corresponding to ca. 11.6–11.3 cal. kyr BP. Somewhat younger records are known from the site of Friesack, Brandenburg (Zeitsufe I) (Gramsch 2000) and Bedburg-Königshoven, Lower Rhineland (Street 1999; Behling & Street 1999) dated to around 11.2–10.7 cal. kyr BP. In Britain the pig is known from the sites of Star Carr, Yorkshire (Fraser & King 1954), Thatcham, Berkshire (King 1962) and Faraday Road, Berkshire (Ellis *et al.* 2003) ranging between ca. 11.5 and 10.6 cal. kyr BP.

Inferred time range. – ca. 11.4 cal. kyr BP to AD 17–1800

Megaloceros giganteus

Giant deer

First appearance datum. – Part of a metatarsus found in a gravel pit at Svenstrup, Ålborg, northern Jylland, directly radiocarbon dated to 31 720 ± 980 ^{14}C yr BP (Ua-2507) (Aaris-Sørensen & Liljegren 2004; Aaris-Sørensen 2006). First appearance *after* LGM is documented by a mandible found by dredging for gravel in Køge Bugt, outside Mosede Havn. The mandible has been directly radiocarbon dated to 12 005 ± 65 ^{14}C yr BP (OxA-10234) (Aaris-Sørensen & Liljegren 2004) corresponding to ca. 13.9–13.8 cal. kyr BP.

Last appearance datum. – A shed antler found at Vævlinge, Bogense, S Fyn, directly radiocarbon dated to 10 700 ± 115 ^{14}C yr BP (K-5658) (Aaris-Sørensen & Liljegren 2004) corresponding to ca. 12.8–12.4 cal. kyr BP (Fig. 19).

Number of dated records. – 11.

Adjacent areas. – The occurrence of the giant deer in Europe during the Weichselian and especially its last northward expansion and abundance during the Late Glacial is well documented (e.g. Lister 1994; Kahlke 1999; von Koenigswald 2002). Four of the seven specimens recorded in Skåne, southern Sweden, have been directly radiocarbon dated and they all fall within the time range for the Danish Post-LGM finds (Aaris-Sørensen & Liljegren 2004).

Inferred time range. – Pre-LGM: Early and Middle Weichselian interstadials. Post-LGM: ca. 14.0–12.4 cal. kyr BP.

Cervus elaphus

Red deer

First appearance datum. – Three specimens are pollen dated to the middle of pollenzone IV (*ex* Jessen 1935)

Fig. 19. Shed antler of giant deer (*Megaloceros giganteus*) found at Vævlinge near Bogense, Fyn. Age: ca. 12.8–12.4 cal. kyr BP. Scale bar = 25 cm. (Photo: Geert Brovad).

corresponding to ca. 10.8 cal. kyr BP and found at Stubberup Mark (Skælskør), Strødam Enge (Hillerød) and Øresømølle (Holbæk) (I. Sørensen, ZMUC files). A direct radiocarbon date of a specimen from Ølene, Bornholm gave 9270 ± 130 ^{14}C yr BP (K-4879), corresponding to ca. 10.6–10.3 cal. kyr BP.

Last appearance datum. – Extant.

Number of dated records. – 411, continuously dispersed between FAD and LAD.

Adjacent areas. – Sommer *et al.* (2008) have compiled directly dated radiocarbon-supported records of European red deer from the late Glacial showing its presence in southern England (Gough's cave) as early as around 15.0 cal. kyr BP, in the German Rheinland (Andernach) around 13.8 cal. kyr BP and in German Lower Saxony (Lemförde) around 13.0 cal. kyr BP. The few available dates from the Younger Dryas cooling are confined to the southern European mainland and southern England. The earliest Post Glacial dates close to Denmark are from two sites in Brandenburg, northern Germany: Friesack dated to around 11.0–10.7 cal. kyr BP (Zeitstufe I) (Gramsch 2000) and Potsdam-Schlaatz with a direct date of 9601 ± 63 ^{14}C yr BP (KIA-9564) corresponding to ca. 11.1–10.8 cal. kyr BP (Benecke *et al.* 2002). Other Early Preboreal remains are known from Star Carr, Yorkshire around 11.0–10.6 cal. kyr BP (Fraser & King 1954) and Thatcham, Berkshire around 11.1–10.7 cal. kyr BP (Hedges *et al.* 1996).

Inferred time range. – ca. 11.4–0 cal. kyr BP.

Alces alces

Elk

First appearance datum. – Complete skeleton found in a bog, Vonsmose, Haderslev, southern Jylland and radiocarbon dated to 11 770 ± 190 ^{14}C yr BP (K-6124) corresponding to ca. 13.8–13.4 cal. kyr BP.

Last appearance datum. – Remains from the Middle Neolithic (Pitted Ware) site of Kainsbakke west of Grenå in eastern Jylland (Richter 1991). The *A. alces* remains are indirectly radiocarbon dated (aurochs) to 4060 ± 50 ^{14}C yr BP (Ua-24709), corresponding to ca. 4.8–4.4 cal. kyr BP.

Number of dated records. – 94. Out of 17 direct radiocarbon dated Late Glacial and Preboreal specimens five are older than Younger Dryas and 12 dates to the Preboreal. The five oldest range from ca. 13.6 to

Fig. 20. Skull belonging to a complete skeleton of an elk (*Alces alces*) found at Kildeskoven, northern Copenhagen. The antlers span 1.65 m. Age: ca. 13.1–12.9 cal. kyr BP. (Photo: Geert Brovad).

12.5 cal. kyr BP (Fig. 20), and the first from the Preboreal is dated to 9920 ± 135 ^{14}C yr BP, corresponding to ca. 11.7–11.2 cal. kyr BP. This gap in the *Alces* record may very well turn out to be real and induced by the cold Younger Dryas period. From the beginning of Preboreal the rest of the records are continuously dispersed till LAD. *Regional* differences can be seen after the Early Atlantic transgressions around 8.0 cal. kyr BP (see the section on Islands).

Adjacent areas. – Important additional data are found in Skåne, southern Sweden. Liljegren & Ekström (1996) reports a date of a metatarsus from Arrie on 12 390 ± 150 ^{14}C yr BP (St-13310). This rather early date has recently been questioned by a new date of the same specimen made in Lund giving an age of 11 345 ± 70 ^{14}C yr BP (LuS-7685) corresponding to ca. 13.3–13.2 cal. kyr BP. Three other direct radiocarbon dates are available from Skåne (Liljegren & Ekström 1996; Larsson *et al.* 2002) of which one also lies before Younger Dryas and the two others in Preboreal. This is in accordance with the Younger Dryas-induced gap seen in the Danish record.

Inferred time range. – ca. 14.0–4.6 cal. kyr BP, most likely with a Younger Dryas Induced Pause ca. 12.6–11.4 cal. kyr BP (for regional differences see the section on Islands).

Rangifer tarandus

Reindeer

First appearance datum. – Antler found in a gravel pit at Lundebjerg south-west of Sæby in northern Jylland – directly radiocarbon dated to 31 910 ± 1315

[14]C yr BP (K-6003) (Aaris-Sørensen 2006). First appearance *after* LGM is documented by a number of bones and antlers found on the Upper Palaeolithic site of Slotseng north of Vojens in southern Jylland with the oldest date of an antler of 12 520 ± 190 ^{14}C yr BP (AAR-906) (Holm & Rieck 1992), corresponding to ca. 14.9–14.3 cal. kyr BP. However, if all the reindeer skeletons found on the site are the result of a single year's autumn hunting, the weighted average of all the ten dated reindeer specimens can be combined to one date giving an average of c 14.2–14.0 cal. kyr BP (Mortensen 2007).

Last appearance datum. – An antler found at Risbanke, Ringsted, Sjælland which has been radiocarbon dated to 9180 ± 80 ^{14}C yr BP (K-7074) (Aaris-Sørensen *et al.* 2007). Three other dates of approximately same age exist, one of them being a worked antler from Nørre Lyngby, northern Jylland yielding an age of 9110 ± 65 ^{14}C yr BP (AAR-8919) (Stensager 2004).

Number of dated records. – 62 (47 direct radiocarbon dates, Aaris-Sørensen *et al.* 2007), *after* LGM continuously dispersed between FAD and LAD (Fig. 5).

Adjacent areas. – The reindeer is well documented from Weichselian deposits all over Europe (e.g. Kahlke 1999; von Koenigswald 2002). The first appearance *after* LGM in Denmark around 12.5 ^{14}C kyr BP is matched by a similar first occurrence in the deglaciated eastern Lithuania around 12.1 (Ukkonen *et al.* 2006) and in western Norway around 12.7–12.2 (Lie 1986, 1990; Mangerud 1977). There seems, however, to be a delay in the immigration of the reindeer into southern Sweden where the oldest reindeer date lies around 11.7 ^{14}C kyr BP (Björck *et al.* 1996) (see Aaris-Sørensen *et al.* 2007 for further discussion).

Inferred time range. – Pre-LGM: Early and Middle Weichselian interstadials. Post-LGM: ca. 14.5–10.3 cal. kyr BP.

Capreolus capreolus

Roe deer

First appearance datum. – A femur found in Køge Bugt, off Solrød Strand, directly radiocarbon dated to 8980 ± 110 ^{14}C yr BP corresponding to ca. 10.2–9.9 cal. kyr BP.

Last appearance datum. – Extant.

Number of dated records. – 284, continuously dispersed between FAD and LAD.

Adjacent areas. – The roe deer was present in the Central Rhineland Neuwied Basin during the Late Glacial Allerød interstadial (sites of Kettig and Miesenheim 2 and 4) (Baales & Street 1996; Baales 2001) and it is recorded in northern Wales with a direct radiocarbon date of 11 795 ± 65 ^{14}C yr BP (OxA-6116) (Hetherington *et al.* 2006). Early Preboreal occurrences are known from the site of Friesack, Brandenburg (Zeitstufe I) (Gramsch 2000) and Bedburg-Königshoven, Lower Rhineland (Street 1999). Early Preboreal records are also known from the two early Mesolithic sites of Thatcham and Faraday Road in Berkshire (King 1962; Ellis *et al.* 2003) and from Star Carr in Yorkshire (Fraser & King 1954).

Inferred time range. – ca. 11.4–0 cal. kyr BP.

Bos primigenius

Aurochs

First appearance datum. – Nine specimens have been directly radiocarbon dated to the very beginning of Preboreal with ages between 9970 and 9810 ^{14}C yr BP (Fig. 21). Geographically they cover the entire Danish area. The oldest specimen, most of a skeleton (♀) found at Store Tåstrup south-west of Holbæk, Sjælland, has a date of 9970 ± 90 ^{14}C yr BP corresponding to ca. 11.6–11.2 cal. kyr BP (unpublished data, P. Gravlund).

Last appearance datum. – Remains of three aurochs specimens, a bull and two cows, excavated at Bimpel east of Tønder, southern Jylland and dated to 2945 ± 25 ^{14}C yr BP corresponding to ca. 3.2–3.1 cal. kyr BP (Aaris-Sørensen 2004). Of the same age is a

Fig. 21. Skull of an aurochs bull (*Bos primigenius*) found in lake deposits at Millinge near Fåborg, Fyn. Greatest span of the horn cores is 114 cm. Age: ca. 11.6–11.2 cal. kyr BP. (Photo: Geert Brovad).

horn found during peat cutting in a bog, Eskær Mose north-west of Frederikshavn, northern Jylland (2810 ± 105 ^{14}C yr BP, K-6121). The horn has been broken ca. 12–14 cm above its base. Nevertheless, it still measures 65 cm along the outer curvature and 10.5 cm across the break. These measurements clearly fall within the range of aurochs and outside the range of domestic oxen (Hatting 1995).

Number of dated records. – 151 of which 42 are direct radiocarbon dates. The records are continuously dispersed between FAD and LAD. *Regional* differences can be seen after the Early Atlantic transgressions around 8.0 cal. kyr BP (see the section on Islands).

Adjacent areas. – The first occurrence of *B. primigenius* in southern Sweden is likewise dated to the very transition between Younger Dryas and Preboreal, whereas the local extinction here takes place as early as ca. 6500 ^{14}C yr BP corresponding to ca. 7.5 cal. kyr BP (Ekström 1993) (see also the section on Islands).

Inferred time range. – ca. 11.4–3.0 cal. kyr BP (for regional differences see the section on Islands).

Bison priscus

Steppe bison

Remarks. – Ten specimens, fragments of horn cores and limb bones, have been found redeposited in glacial and glaciofluvial deposits in Jylland, Fyn and Sjælland. Radiocarbon dating has been tried twice giving a non-finite date of >43 000 ^{14}C yr BP (Rolsted, south-east of Odense, Fyn, AAR-2708) and an apparent age of 45 600 ± 2000 ^{14}C BP (Grønninghoved, south of Kolding, south-east Jylland (LuS-6142)) (Aaris-Sørensen 2006).

Adjacent areas. – Widely distributed and abundant all over Europe during the Weichselian (see, e.g. Kahlke 1999; von Koenigswald 2002).

Inferred time range. – Early and Middle Weichselian interstadials.

Bison bonasus

European bison

First appearance datum. – Skull fragments, vertebrae and pelvis of a bison found during peat cutting in Jarmsted Mose at Brovst in northern Jylland, radiocarbon dated to 10 000 ± 80 ^{14}C yr BP (AAR-4544) corresponding to ca. 11.7–11.4 cal. kyr BP.

Last appearance datum. – Ribs and thoracic vertebrae (with complete spinous processes characteristic of bison) found during drainage in a bog at Akkerup between Assens and Fåborg, SW Fyn, radiocarbon dated to 9540 ± 85 ^{14}C yr BP (K-6005) corresponding to ca. 11.0–10.7 cal. kyr BP. Three much younger specimens have been recorded (dated to the Neolithic/Bronze Age transition, Iron Age and the medieval time) – these are discoveries from bogs of two skulls with horn cores (trophy/ sacrifice?) and a metacarpal from the moat of a medieval castle. These remains probably point to a North European Subboreal/Subatlantic bison population not so far from Denmark.

Number of dated records. – 8.

Adjacent areas. – Eight specimens from southern Sweden have been radiocarbon dated ranging from ca. 11.1–10.2 cal. kyr BP (Liljegren & Ekström 1996) and from the Upper Palaeolithic site of Stellmoor two bisons have recently been radiocarbon dated to 10 070 ± 50 ^{14}C yr BP (KIA-3331) (Benecke 2000) and 8970 ± 75 ^{14}C yr BP (OxA-3628) (Bratlund 1999). If calibrated into calendar years the German dates cover the range between ca. 11.7 and 10.0 cal. kyr BP. Seen in the light of the Danish and the Swedish dates this could very well cover the Early Holocene occurrence of the wisent in northern Europe.

Inferred time range. – ca. 11.4–10.0 cal. kyr BP.

Saiga tatarica

Saiga

First and last appearance data. – Only one single record, a specimen found embedded in till near Boltinggaards Skov, south of Odense, Fyn and dated twice to 13 880 ± 140 and 14 040 ± 200 ^{14}C yr BP (AAR-1456 and 1977), corresponding to ca. 17.2– 16.4 cal. kyr BP (Aaris-Sørensen *et al.* 1999) (Fig. 22).

Adjacent areas. – The single Danish saiga specimen fits into a well-documented immigration wave of saiga into Late Glacial Europe around 15.0–12.0 ^{14}C yr BP (ca. 18.0–14.0 cal. kyr BP). This expansion reached south-west France (Delpech 1983), south-east France (Crégut-Bonnoure & Gagnière 1981; Crégut-Bonnoure 1992), south England (Currant 1987), Germany (Kahlke 1990, 1992) and as far north as Denmark (Aaris-Sørensen *et al.* 1999). The occurrence of saiga in Europe during the Late Weichselian indicates the presence of a dry, continental steppe environment (see also *D. moschata, O. pusilla, S. major* and

Fig. 22. Left part of a skull and horn core of a saiga (*Saiga tatarica*) found embedded in till at Boltingsgaards Skov south of Odense, Fyn. Age: ca. 17.2–16.4 cal. kyr BP. (Photo: Geert Brovad). Photo is 2/3 of natural size.

M. gregalis) which included southern Scandinavia at least around 17.2–16.4 cal. kyr BP.

Inferred time range. – ca. 17.2–14.0 cal. kyr BP

Ovibos moschatus

Musk ox

Remarks. – Only two specimens, both skulls, have been found in Danish glacial deposits, one in a gravel pit at Romalt, Randers, E. Jylland and another in a gravel pit at Bannebjerg, Helsinge, N Sjælland. The latter has been radiocarbon dated to 28 490 ± 350 ^{14}C yr BP (AAR-4188) (Aaris-Sørensen 2006) (Fig. 4).

Adjacent areas. – Fossils are known from Weichselian deposits all over Europe and the musk ox seems to have been most abundant during the coldest and driest periods (Kahlke 1999; Raufuss & von Koenigswald

1999). Besides the two Danish examples, a few other Scandinavian remains of musk ox have been detected in Norway and Sweden. Two well-preserved vertebrae were found in gravel at Indset, Dovre, Norway and a tentative radiocarbon dating gave a non-finite date of >40 000 ^{14}C yr BP (Hufthammer 2001). Three finds have been reported from Sweden including a tibia fragment from a gravel pit at Nol, Dösebacka, Bohuslän (Munthe 1905) and a skull and a humerus fragment from gravel pits at Åskott and Frösön, both near Östersund in Jämtland (Borgen 1979). The Nol specimen has, according to Liljegren & Lagerås (1993), been radiocarbon dated to ca. 32.0 ^{14}C kyr BP.

Inferred time range. – Early and Middle Weichselian interstadials.

Cetacea

Lagenorhynchus albirostris

White-beaked dolphin

First appearance datum. – Skull found in *Cardium* clay at Gniben, Sjællands Odde and direct radiocarbon dated to 6530 ± 110 ^{14}C yr BP (K-6767); corrected for marine reservoir effect, −400 years, this corresponds to ca. 7.6–7.3 cal. kyr BP (Aaris-Sørensen *et al.* in press) (Fig. 23).

Last appearance datum. – Extant.

Number of dated records. – 10, dispersed between FAD and LAD with gaps in Early Subboreal and Early Subatlantic.

Adjacent areas and recent evidence. – The modern distribution points at a much earlier first appearance than documented by the subfossil record. This assumption is supported by a single Swedish find from a shell mound near Uddevalla. Other vertebrate remains and shells from this locality have given dates between 10 000 and 11 000 ^{14}C yr BP (Lepiksaar 1966; Fredén 1984). The species is common in the Danish waters in recent times and believed to breed in the Danish part of the North Sea and Skagerrak (Kinze 2007) and has probably been part of the Danish fauna continuously throughout Late and Post Glacial.

Inferred time range. – Pre-LGM: whenever a marine palaeo-Kattegat–Skagerrak existed. Post-LGM: ca. 18.0–0 cal. kyr BP.

Delphinus delphis

Common dolphin

First appearance datum. – Only two subfossil records of common dolphin in all – both specimens found on Mesolithic sites in Jylland (Møllegården, Vilsted, south of Løgstør and Dyrholmen, Randers) belonging to Ertebølle–Early Neolithic cultures which date to around 7.0–5.5 cal. kyr BP.

Last appearance datum. – Extant, generally rare but periodically rather common.

Number of dated records. – 2.

Recent evidence. – The common dolphin is a southern continental shelf species and today it visits the inner Danish waters occasionally in connection with increasing flow of Atlantic waters into the Baltic

Fig. 23. Skull of white-beaked dolphin (*Lagenorhynchus albirostris*) found at Gniben, Sjællands Odde. Age: ca. 7.6–7.3 cal. kyr BP. Scale bar = 10 cm. (Photo: Geert Brovad).

(Kinze 2007). These invasions of small schools of dolphins have probably taken place since the Early Atlantic transgressions some 8000 years ago.

Inferred time range. – ca. 8.0–0 cal. kyr BP.

Tursiops truncatus
Bottlenose dolphin

First appearance datum. – Several specimens found on Ertebølle sites in northern and eastern Jylland, Fyn and northern Sjælland. The oldest come from the site of Lystrup Enge, Århus (I. B. Enghoff, unpublished manuscript) which date back to around 7.5 cal. kyr BP.

Last appearance datum. – Extant, rare, occurs periodically in both inner and offshore Danish waters (Kinze 2007).

Number of dated records. – 16.

Recent evidence. – In recent times the species is common in the southernmost part of the North Sea and southwards from here. Occasionally, it spreads northwards into the North Sea either directly or via the Irish Sea and north of Scotland (Kinze 2007). These expansions also reach Danish waters from time to time and this has probably happened ever since the Early Atlantic transgressions some 8000 years ago.

Inferred time range. – ca. 8.0–0 cal. kyr BP.

Stenella sp.
Stenella dolphin

First appearance datum. – Only one single bone (the last thoracic or first lumbar vertebra) has been found of a dolphin belonging to the genus *Stenella*. Species identification is not possible based on morphological characters alone. The vertebra was excavated on the submarine Ertebølle site of Tybrind Vig between Middelfart and Assens, north-west Fyn (Trolle-Lassen 1985). The occupation of the site covers the whole Ertebølle period corresponding to ca. 7.4–5.9 cal. kyr BP.

Last appearance datum. – Extant, a rare visitor.

Number of dated records. – 1.

Recent evidence. – In recent years, the striped dolphin, *Stenella coeruleoalba*, has been an occasional visitor from more southerly and warmer waters (Kinze 2007). Most likely these visits into Danish waters have

taken place since the Early Atlantic transgressions and probably with a peak during the Holocene thermal maximum between ca. 7.0 and 4.0 cal. kyr BP. The Stenella dolphin from Tybrind Vig might therefore represent one of these early striped dolphin visits.

Inferred time range. – ca. 8.0–0 cal. kyr BP.

Orcinus orca
Killer whale

First appearance datum. – A directly radiocarbon-dated specimen from the Ertebølle site Lystrup Enge, Århus with an age of 6800 ± 115 ^{14}C yr BP (corrected for marine reservoir effect, −400 years). Calibrated, this corresponds to ca. 7.8–7.6 cal. kyr BP (Aaris-Sørensen et al. in press).

Last appearance datum. – Extant, quite common visitor.

Number of dated records. – 21, continuously dispersed between FAD and LAD.

Adjacent areas and recent evidence. – Two specimens found in a Late Glacial/Early Holocene shell mound at Otterö in Bohuslän, Sweden (Lepiksaar 1966) and in Younger Yoldia Clay near Langholt, WSW of Sæby in northern Jylland (Nordmann 1944; Møhl 1971a) suggest a much earlier occurrence of *O. orca* in Danish waters. This is in accordance with the current distribution of the species which encompasses all oceans from Arctic to Antarctic waters and it makes it probable that it has visited Danish waters ever since the Middle Weichselian.

Inferred time range. – Pre-LGM: whenever a marine palaeo-Kattegat–Skagerrak existed. Post-LGM: ca. 18.0–0 cal. kyr BP.

Phocoena phocoena
Harbour porpoise

First appearance datum. – Remains found on four Mesolithic sites on Sjælland dated to the Kongemose/Ertebølle culture (Bloksbjerg, Carstensminde, Vedbæk Boldbaner, Villingebæk Ø (for a closer localization, see *H. grypus*)) which falls within ca. 8.0–6.0 cal. kyr BP.

Last appearance datum. – Extant.

Number of dated records. – 49, continuously dispersed between FAD and LAD.

Adjacent areas. – A much earlier occurrence in the Kattegat area is documented by a directly radiocarbon-dated specimen found in marine clay at Kyvik on the coast of Halland (Lepiksaar 1966; Fredén 1984). The date of 10 630 ± 350 ^{14}C yr BP falls within 12.9–12.0 cal. kyr BP indicating the presence of *P. phocoena* in Skagerrak/Kattegat ever since the Late Glacial followed by an expansion into the inner Danish waters and the Baltic in the wake of the Littorina transgressions.

Inferred time range. – ca. 18.0–0 cal. kyr BP.

Delphinapterus leucas/Monodon monoceros

Beluga whale/Narwhale

Remarks. – Only seven specimens (atlas and caudal vertebrae) of Monodontidae whales have been found in Denmark. Ancient DNA analyses have identified two specimens as *M. monoceros* (E. Garde, personal communication 2009). The rest cannot be identified with certainty either by morphology or DNA. However, several complete or almost complete skeletons of identifiable *D. leucas* specimens have been found on the Swedish west coast (Lepiksaar 1966; Fredén 1984) so both species seem to have been present in south Scandinavia.

First appearance datum. – A vertebra found in a gravel pit at Åsbakken, Hjørring has an infinite radiocarbon date of >38 000 ^{14}C yr BP (K-7103) and two other finds from gravel pits in northern and eastern Jylland have been radiocarbon dated to 33 140 ± 1460 ^{14}C yr BP (K-7107) and 32 220 ± 1480 ^{14}C yr BP (K-7102). First occurrence *after* LGM is dated by a vertebra found near Fredrikshavn in northern Jylland with an age of 12 000 ± 170 ^{14}C yr BP (K-7104) corresponding to ca. 14.0 cal. kyr BP (Aaris-Sørensen *et al.* in press).

Last appearance datum. – An atlas found on the Mesolithic site of Meilgaard, eastern Jylland, ca. 7–5 cal. kyr BP (Winge 1899).

Number of dated records. – 5.

Adjacent areas. – 10 radiocarbon-dated specimens known from the Swedish west coast ranging from ca. 12 000 to 8300 ^{14}C yr BP (Fredén 1984; Aaris-Sørensen *et al.* in press) corresponding to ca. 14–9 cal. kyr BP.

Inferred time range. – Pre-LGM: whenever a marine palaeo-Kattegat–Skagerrak existed. Post-LGM: 18.0–

11.7 cal. kyr BP (common) and ca. 11.7–0 cal. kyr BP (very rare visitor).

Hyperoodon ampullatus

Northern bottlenose whale

Remarks. – Only one single subfossil record exists of this whale in Denmark. The specimen, a cervical vertebra, has not yet been dated, but it was found in a depth of ca. 1.5 m during drainage about 150 m from the recent coast line at Dragsholm Hovedmark, Fårevejle in NW Sjælland. Presumably the vertebra represents a stranding of a bottlenose whale once on the former Littorina Sea coast line. A single subfossil find is also reported from the Swedish west coast near Uddevalla, found in clay and believed to date to the Pleistocene/Holocene transition (Lepiksaar 1966, 1986).

Today the species occurs exclusively in the North Atlantic but stray visitors are seen in Danish–Swedish common waters from time to time (Kinze 2007). These rare visits have probably occurred ever since the LGM (ca. 18.0–0 cal. kyr BP).

Physeter macrocephalus

Sperm whale

First appearance datum. – A single specimen found on the Mesolithic Ertebølle site of Vængesø III, Helgenæs, eastern Jylland. Several radiocarbon dates of other mammalian remains place the occupation of this site between ca. 6.8 and 5.8 cal. kyr BP (I. B. Enghof unpublished).

Last appearance datum. – Extant.

Number of dated records. – 6, besides the above-mentioned Late Atlantic specimen three belong to the Late Subboreal (Younger Bronze Age) and two to the Late Subatlantic (Historical times).

Recent evidence. – The sperm whale is an oceanic and cosmopolitan species which can be found in all oceans from the Arctic to the Antarctic waters. At present it is rather frequently encountered in Denmark during stranding events along the west coast of Jylland (Kinze 2007). This may have happened as full marine conditions were established in the North Sea and the English Channel around 8.0 cal. kyr BP.

Inferred time range. – Pre-LGM: whenever a marine palaeo-Kattegat–Skagerrak existed. Post-LGM: ca. 18.0–0 cal. kyr BP.

Balaenoptera acutorostrata

Common minke whale

First appearance datum. – A caudal vertebra found in a clay pit at Lindholm, Nørre Sundby in northern Jylland has been radiocarbon dated to 29 320 ± 1240 ^{14}C yr BP (K-6771) showing it as a member of the Older Yoldia Sea fauna (Aaris-Sørensen *et al.*, in press).

Last appearance datum. – Extant.

Number of dated records. – 4. Besides the Middle Weichselian specimen mentioned above, three others belong to the Post Glacial. Two of them have been radiocarbon dated to 5130 ± 100 ^{14}C yr BP (K-6803) (Holmeenge I, Randers) and 3540 ± 90 ^{14}C yr BP (K-6770) (Nordby Hede, Samsø, part of skull identified as *Balaenoptera* cf. *acutorostrata*) which covers the period of ca. 6.0–3.7 cal. kyr BP (Aaris-Sørensen *et al.* in press). The third was found in Asaa, Hjørring in layers dated to the Middle Ages (AD 13–1500).

Recent evidence. – Today the species occurs in the entire North Atlantic. It is the most common baleen whale in Danish waters and considered a coastal species (Kinze 2007). Despite the very few, but chronologically well spread, findings it seems reasonable to believe that the minke whale has been a frequent guest in the Danish waters before as well as after the LGM.

Inferred time range. – Pre-LGM: whenever a marine palaeo-Kattegat–Skagerrak existed. Post-LGM: ca. 18.0–0 cal. kyr BP.

Balaenoptera physalus

Fin whale

First appearance datum. – Large part of a skeleton found on a raised shore line belonging to the Littorina Sea at Åsted, Salling in N Jylland and dated to 5060 ± 95 ^{14}C yr BP (K-5998), which corresponds to ca. 5.9–5.7 cal. kyr BP (Stenstrop 1994).

Last appearance datum. – Extant.

Number of dated records. – 4. Besides the above-mentioned Middle Atlantic specimen one was found at Vængesø, Helgenæs, eastern Jylland and dated to 2890 ± 80 ^{14}C yr BP (K-5661) (ca. 3.2–2.9 cal. kyr BP) (Aaris-Sørensen *et al.* in press), another at the

Pre-Roman Iron Age site Smedegård, north of Thisted in northern Jylland (ca. 2.0 cal. kyr BP) (Raahauge 2002) and a third at Egense on the coast of Kattegat east of Ålborg, dated to 1250 ± 45 ^{14}C yr BP (LuS-7413) corresponding to ca. AD 680–850.

Recent evidence. – Today the fin whale is an oceanic species found in all oceans from the arctic to the Antarctic. Individuals from the North Atlantic periodically frequent the Danish coasts especially the deep fjords of the east coast of Jylland (Kinze 2007) – this has been possible for the last ca. 8000 years.

Inferred time range. – At least from ca. 8.0 cal. kyr BP (probably much earlier) to 0.

Balaenoptera cf. *musculus*

Blue whale

Remarks. – Out of seven specimens ascribed to this species only one has been radiocarbon dated: an occipital part of a skull found during ploughing on the reclaimed Tastum Sø south of Skive. During the Littorina Sea period Tastum Sø formed the southernmost part of Skive Fjord. The morphology of the fragment conform with *B. musculus* and *B. physalus*, but the large size make *musculus* the best match. The skull has been radiocarbon dated to 5160 ± 90 ^{14}C yr BP (K-6797) corresponding to ca. 5.7–6.0 cal. kyr BP. The occurrence in Danish waters during historical times is extremely rare and the species has not been documented in Denmark since 1936 (Kinze 2007).

Megaptera novaeangliae

Humpback whale

Remarks. – Only two dated subfossil finds of this species are known from Denmark. A rib has been found at Vester Tværsted, east of Hirtshals in northern Jylland and dated to 12 100 ± 185 ^{14}C yr BP (ca. 14.2–13.7 cal. kyr BP) (Aaris-Sørensen *et al.* in press). The morphology of the rib conform well with the humback whale but other species cannot be totally excluded and it should therefore be assigned to *M.* cf. *novaeanglia*. The other specimen, a vertebra found at Søborg Slotsruin south of Gilleleje in northern Sjælland, can, however, be assigned to *M. novaeanglia* with certainty (Degerbøl 1946b). The vertebra is dated to 705 ± 75 ^{14}C yr BP corresponding to ca. AD 1230–1390 (Aaris-Sørensen *et al.* in press). Today individuals from the recovered North Atlantic population are occasionally sighted in inner Danish waters (Kinze 2007).

Eubalaena glacialis

Atlantic northern right whale

Remarks. – There are only two dated subfossil finds of this whale in Denmark. A lumbar vertebra was found in a gravel pit at Staurby Skov, Middelfart, NW Fyn and dated to 17 830 ± 590 ^{14}C yr BP (K-6768) (ca. 22.0–20.5 cal. kyr BP) and most of a left flipper (radius, ulna, metacarpal I + II + III) was found on an old shore line at Ballerum north-east of Thisted in northern Jylland and dated to 6410 ± 110 ^{14}C yr BP (K-6213) (ca. 7.4–7.2 cal. kyr BP) (Aaris-Sørensen *et al.* in press). All remains can be clearly distinguished from *Balaena mysticetus*. The northern right whale is a slow whale, easy to kill and easy to tow as it floats well because of a very thick layer of blubber and so it has been exposed to a severe decimation in historical times. In Denmark it has only been documented once in Vejle Fjord 1838 (Kinze 2007). Nevertheless, the two subfossil finds make it probable that the species, prior to the modern exploitation, visited the Danish waters occasionally ever since the Early Weichselian.

Balaena mysticetus

Greenland right whale/bowhead whale

First appearance datum. – A rib found in Yoldia Clay at Ravnsholt near Sæby (Winge 1904) and dated to 14 110 ± 215 ^{14}C yr BP (K-7110) corresponding to ca. 17.2–16.5 cal. kyr BP (Aaris-Sørensen *et al.* in press).

Last appearance datum. – Radius found at Strandby north of Frederikshavn in northern Jylland and dated to 1500 ± 65 ^{14}C yr BP (K-6820) corresponding to ca. AD 440–640 (Aaris-Sørensen *et al.* in press).

Number of dated records. – 8; five fall within 14.0–11.6 ^{14}C kyr BP and three within 4.3–1.5 ^{14}C kyr BP, all remains found in northern Jylland.

Adjacent areas. – Around 30 finds are recorded on the Swedish west coast and 17 have been radiocarbon dated giving 15 Late Glacial dates ranging from 12.6 to 10.2 ^{14}C kyr BP as well as two younger dates of 6.4 and 4.2 ^{14}C kyr BP (Fredén 1984; Aaris-Sørensen *et al.* in press). Like the other 'right whale' (*E. glacialis*) the bowhead whale is a slow swimmer and floats well after death and it has therefore likewise been almost wiped out by commercial whaling in historical times. This circum-arctic whale is seldom sighted south of 45° north latitude and there are no Danish records of the species from historical times.

Inferred time range. – Pre-LGM: whenever a marine palaeo-Kattegat–Skagerrak existed. Post-LGM: ca. 18.0–11.7 (common) and ca. 11.7–0 cal. kyr BP (very rare visitor).

Balaena mysticetus/Eubalaena glacialis

Right whales

Remarks. – Remains of seven right whales which defy a clear species identification on the basis of osteological characters alone have, nevertheless, been radiocarbon dated as they can provide important information on glacial history and shore line displacement. Like the *B. mysticetus* specimens they all come from northern Jylland and the dates also fall into two groups with five Late Glacial (13.7–11.3 ^{14}C kyr BP) and two young dates (4.6 and 1.4 ^{14}C kyr BP) (Aaris-Sørensen *et al.* in press).

The terrestrial fauna prior to LGM, ca. 115–22 kyr BP

As it can be seen from an examination of the preceding Systematic section remains of only nine terrestrial species have been recorded from the Early and Middle Weichselian prior to LGM. Except for two species of lemmings, *L. lemmus* and *D. torquatus*, of which teeth and bones have been found in lacustrine sediments in a coastal cliff on the island of Møn (Bennike *et al.* 1994) the others are large mammals, namely *M. primigenius*, *E. ferus*, cf. *C. antiquitatis*, *M. giganteus*, *R. tarandus*, *B. priscus* and *O. moschatus*. It is common to these members of the last glacial megafauna that their remains have been found redeposited in glacial and glaciofluvial deposits and all show more or less clear signs of having been transported by ice or water. With only about 180 bones, teeth and antlers recovered in total the remains are few in number compared with the abundance seen in central Europe. They are, nevertheless, interesting as they provide important information on the interaction of glaciation chronology and environmental history in the region including the mechanisms controlling displacement of species and faunas.

These interactions have recently been analysed by Aaris-Sørensen (2006) and it could be concluded that the range extensions and contractions of the mammalian species in no case seem to be in conflict with the glacial history and palaeogeography as outlined by recent geological studies (e.g. Houmark-Nielsen & Kjær 2003). On the contrary, the megafaunal data support these scenarios which are reconstructed on the basis of calibrated radiocarbon ages from terrestrial plant macrofossils, marine molluscs and for-aminifera combined with optically stimulated luminescence (OSL) dates of lacustrine and marine sediments. Moreover, the interrelationship between glacial history and the faunal changes observed supports the idea that the displacement of a species' distribution is best explained as a combination of expansions and local extinctions of marginal populations. During times of climatic amelioration a northward expansion can be seen of the central European megafauna into the newly deglaciated areas in south Scandinavia and it seems that the rate of this expansion of large herbivorous mammal populations was controlled by the rate of 'habitat migration' – the time lag in vegetational response to the climatic improvement. Deterioration of the ecological conditions, on the other hand, led to local extinction of the marginal, northernmost Scandinavian populations and hereby a 'contraction' of the geographical range. As stressed by von Koenigswald (2003) emigration, in the true sense of the word, would normally not be possible, as it would require uninhabited favourable areas. With a 'mammoth steppe fauna' well-established right south

of the south Scandinavian region these conditions were not at hand.

As these conclusions were drawn, new dates of mammoths and rhinoceros have been achieved. Table 1 and Figure 24 update the number of species and specimens and present their locations and dates. The number of dated specimens has increased to 32 and geographically they cover Jylland, Als, Fyn, Lolland, Sjælland and Skåne.

Two dates of rhinoceros are now included. As mentioned previously, only one of the five recovered specimens can be identified with certainty to *C. antiquitatis*. As this specific specimen is unavailable for dating an attempt has been made to date two limb bones found in two gravel pits near Seest in SE Jylland. A finite radiocarbon date would in all probablity place the rhinoceros within the Middle–Late Weichselian making an ecological identification to the cold-adapted *C. antiquitatis* most likely. Unfortunately, one of the dates came out as a non-finite date older than 44 000 ^{14}C yr BP (LuS-7375) and the other as an 'apparent age' of 41 500 ± 1800 ^{14}C yr BP (LuS-7377). Whether the latter is a true

Table 1. Radiocarbon dates of Weichselian megafauna remains.

Species		Locality	Lab. no.	Sample type	^{14}C age (BP)	δ^{13}C (‰ PDB)	Reference
Megaloceros giganteus	1	Svenstrup, NE Jylland	Ua-2507	Bone	31 720 ± 980	−21.0	1
Rangifer tarandus	2	Lundebjerg, N Jylland	K-6003	Antler	31 910 ± 1315	−17.9	
Bison priscus	3	Rolsted, Fyn	AAR-2708	Bone	>43 000	−22.8	
B. priscus	4	Grønninghoved, SE Jylland	LuS-6142	Bone	45 600 ± 2000		
Saiga tatarica	5	Boltinggårds Skov, Fyn	AAR-1977	Bone	14 040 ± 200	−18.7	3
Ovibos moschatus	6	Bannebjerg, N Sjælland	AAR-4188	Bone	28 490 ± 350	−18.7	
Equus ferus	7	Vejrhøj, SE Jylland	KiA-19280	Bone	38 990 + 1100/−970	−19.9	
cf. *Coelodonta antiquitatis*	8	Seest, SE Jylland	LuS-7375	Bone	>44 000		
cf. *C. antiquitatis*	9	Seest, SE Jylland	LuS-7377	Bone	41 500 ± 1800		
Mammuthus primigenius	10	Rosmos, E Jylland	K-4587	Molar	>39 600	−21.1	2
M. primigenius	11	Sønder Kollemorten, E Jylland	K-4191	Tusk	>37 900	−20.2	2
M. primigenius	12	Rolsøgård, E Jylland	OxA-1090	Molar	41 500 ± 700	−20.5	
M. primigenius	13	Arrie, Skåne	LuS-6651	Molar	40 200 ± 800		6
M. primigenius	14	Nymølle, Sjælland	K-6000	Molar	34 640 ± 1830	−19.5	2
M. primigenius	15	Örsjö, Skåne	Lus-6342	Tusk	34 500 + 400		6
M. primigenius	16	Bårslöv, Skåne	OxA-10193	Molar	33 850 ± 700	−20.0	
M. primigenius	17	Rønninge, Fyn	LuS-6571	Tusk	32 650 ± 270	−21.4	
M. primigenius	18	Lundebjerg, N Jylland	K-4190	Molar	32 460 + 970/−870	−20.2	2
M. primigenius	19	Kiskelund, SE Jylland	K-3696	Tusk	31 840 + 1010/−870	−20.5	2
M. primigenius	20	Saxkøbing, Lolland	K-3807	Tusk	29 570 + 950/−870	−19.6	2
M. primigenius	21	Stengårdens Grusgrav, Sjæll.	K-4192	Tusk	27 810 ± 610	−20.0	2
M. primigenius	22	Høgebjerg, Als	LuS-7416	Molar	26 400 ± 200		4
M. primigenius	23	Djurslöv, Skåne	LuS-6336	Molar	26 150 ± 200		6
M. primigenius	24	Nymølle, Sjælland	K-5999	Molar	26 060 ± 1070	−19.5	
M. primigenius	25	Ny Stengård, Sjælland	K-3805	Tusk	25 760 + 840/−770	−21.2	2
M. primigenius	26	Østrupgård, Fyn	K-3809	Tusk	25 480 + 560/−520	−20.4	2
M. primigenius	27	Hadsund, NE Jylland	K-3699	Tusk	25 110 ± 440	−19.5	2
M. primigenius	28	Munke Bjergby, Sjælland	K-3806	Tusk	24 190 ± 420	−20.9	2
M. primigenius	29	Hornbæk, N Sjælland	LuS-7414	Molar	22 900 ± 150		4
M. primigenius	30	Myrup Banke, Sjælland	K-3703	Bone	21 530 ± 430	−20.8	2
M. primigenius	31	Højballegård, E Jylland	LuS-7415	Molar	19 940 ± 120		4
M. primigenius	32	Lockarp, Skåne	Lu-796	Tusk	13 360 ± 95		5

Locality numbers (1–32) refer to numbers used in the text and in Figure 24. References: 1 = Aaris-Sørensen & Liljegren 2004; 2 = Aaris-Sørensen *et al.* 1990; 3 = Aaris-Sørensen *et al.* 1999; 4 = Aarppe & Karhu, in press; 5 = Berglund *et al.* 1976. (Revised after Aaris-Sørensen 2006).

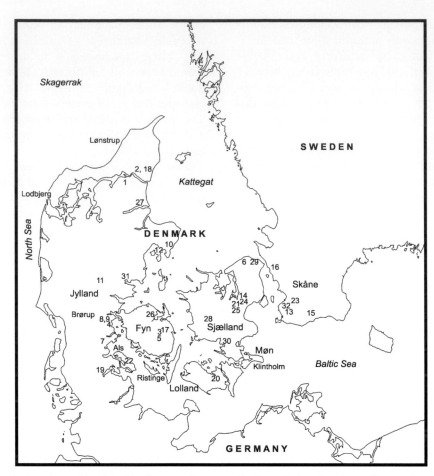

Fig. 24. Map showing the locations of dated Weichselian megafauna remains in south Scandinavia. Locality numbers (1–32) refer to the text and to Table 1. (Revised after Aaris-Sørensen 2006).

finite date or the result of a slight contamination of an infinite sample with modern carbon is hard to tell. Nevertheless, taking into account the wide distribution and abundance of *C. antiquitatis* on the rest of the Weichselian mammoth steppe, I include the species as 'most likely present' in southern Scandinavia as well.

Table 2 places the dated species and specimens in chronological order and in relation to the three Weichselian stadials recognized in southern Scandinavia, the Ristinge, Klintholm and Jylland stadials. The infinite dates of rhinoceros, mammoths and steppe bisons, which probably includes the bison date of 45.6 ^{14}C kyr BP (Table 1, no. 4), might very well represent an interstadial fauna from the time before the Ristinge ice advance. Less likely, they represent dates from the very beginning of the post-Ristinge interstadial and hereby add to the finite dates of mammoth and rhinoceros on 41.5 ^{14}C kyr BP.

In Figure 25 the dated specimens have been plotted according to the regional chronostratigraphy *and* their geographical dispersal. Chronologically they range between ca. 45 and 20 ^{14}C kyr BP, a period

with prevailing interstadial conditions correlated with the Hengelo and Denekamp Interstadial of northwestern Europe. With some modifications throughout the period, the ice margin more or less followed the Norwegian coastline and continued along a NW–SE line from the Oslo Fiord area to the island of Öland in the southern Baltic (Houmark-Nielsen & Kjær 2003). To give a rough estimate of the distance between the dated specimens and the ice margin, the dates have been plotted on a line running parallel with the ice margin and then displaced from SW to NE of the region. This means that dates standing on the same vertical line share the same distance to the ice margin with the Kiskelund mammoth (Table 1, no. 19) being farthest away with about 300–320 km to the ice margin and the Scanian mammoths at Örsjö and Bårslöv (Table 1, nos 15, 16) being closest with only about 30–50 km to the glacier. Reading the chart horizontally, on the other hand, it can be seen, e.g. that around 25 ^{14}C kyr BP the mammoth was widely dispersed from NE and SE Jylland and eastwards over Fyn and Sjælland to Skåne (Table 1, nos. 22–28).

Table 2. Dates for megafauna in chronological order and seen in relation to the Ristinge, Klintholm and Jylland stadial.

Ristinge stadial	
Interstadial, pre- or post-Ristinge	
>44.0	*Coelodonta antiquitatis,* Jylland
>43.0	*Bison priscus,* Jylland
>39.6	*Mammuthus primigenius,* Jylland
>37.9	*M. primigenius,* Jylland
45.6	*B. priscus,* Jylland (infinite?)
Interstadial, post-Ristinge	
41.5	*C. antiquitatis,* Jylland
41.5	*M. primigenius,* Jylland
40.0	*M. primigenius,* Skåne
39.0	*E. ferus,* Jylland
34.5	*M. primigenius,* Sjælland
34.5	*M. primigenius,* Skåne
34.0	*M. primigenius,* Skåne
Klintholm stadial	
32.5	*M. primigenius,* Fyn
32.5	*M. primigenius,* Jylland
32.0	*M. primigenius,* Jylland
32.0	*Rangifer tarandus,* Jylland
31.5	*Megaloceros giganteus,* Jylland
Interstadial, post-Klintholm	
29.5	*M. primigenius,* Lolland
28.5	*Ovibos moschatus,* Sjælland
28.0	*M. primigenius,* Sjælland
26.5	*M. primigenius,* Als
26.0	*M. primigenius,* Sjælland
26.0	*M. primigenius,* Sjælland
26.0	*M. primigenius,* Skåne
25.5	*M. primigenius,* Fyn
25.0	*M. primigenius,* Jylland
24.0	*M. primigenius,* Sjælland
23.0	*M. primigenius,* Sjælland
21.5	*M. primigenius,* Sjælland
20.0	*M. primigenius,* Jylland
Jylland stadial	
Earliest post-Jylland dates	
14.0	*Saiga tatarica,* Fyn
13.5	*M. primigenius,* Skåne

Ages are un-calibrated ^{14}C kyr BP. (Revised after Aaris-Sørensen 2006).

Recently, Ukkonen *et al.* (2007) have focused on the apparent discrepancy between a Middle Weichselian glacial advance in the Baltic depression, the Klintholm advance, reaching south-eastern Denmark and northern Germany (Houmark-Nielsen & Kjær 2003; Houmark-Nielsen 2004) and the contemporaneous occurrences of a megafauna in most of the circum-Baltic region. Apparently, the Danish OSL-based glaciation chronology in the Late–Middle Weichselian does not leave time and space for ice-free conditions suitable for a megafuana in Sweden. Whether this discrepancy is due to uncertainties linked to calendar time-scale conversions of radiocarbon ages, to OSL datings or to the fact that we simply need to consider a much more dynamic behaviour of the Scandinavian Ice Sheet should not be discussed here (see Ukkonen *et al.* 2007). Here, it is important to emphasize that the Danish megafaunal assemblage still supports a glacial history where MIS 3 glaciers flowed through the Baltic depression at least twice, during the Ristinge

and the Klintholm advances. The dynamics of the megafauna as shown in Figure 25 and Table 2 match the advances and retreats of the ice cap by subsequent local extinctions and expansions (Aaris-Sørensen 2006).

The seven large herbivorous mammals together with the two small lemmings presented here are just a mere shadow of the contemporaneous central European fauna belonging to the last glacial period (e.g. Kahlke 1999; von Koenigswald 2002). Given the repeated glaciations in the area and the resulting destruction and redeposition of bone remains, only few of the largest and most resistant parts of the skeleton have a chance of being retrieved. The chances increase considerably *after* the last glacial advance where some of the earliest appearing mammals undoubtedly represent a last northward expansion of the European glacial fauna. This applies for a few remains found in sediments belonging to one of several stillstands and readvances of the ice during the final deglaciation, namely remains of *S. tatarica* dated to ca. 14.0 cal. kyr BP (Table 1, no. 5) (Aaris-Sørensen *et al.* 1999) and *M. primigenius* dated to ca. 13.3 cal. kyr BP (Table 1, no. 32) (Berglund *et al.* 1976). However, small mammals such as *L. timidus, M. oeconomus* and *M. gregalis* found *in situ* in slightly younger deposits may also be included in the Weichselian fauna together with carnivores such as *C. lupus, V. lagopus, U. maritimus, M. erminea, M. nivalis* and *G. gulo.* Together the fossil and the inferred evidence show that the European mammoth steppe fauna expanded into an ice-free southern Scandinavia during all Weichselian interstadials.

The terrestrial fauna after LGM, ca. 17–0 kyr BP

After long Early and Middle Weichselian periods with prevailing interstadial conditions the whole of south Scandinavia was finally glaciated in early Late Weichselian. The Scandinavian Ice Sheet reached its maximum position in the area around 22 kyr BP with the south-western margin running along the Main Stationary Line in Mid-Jylland (Fig. 2F). Outside the ice margin, in present-day western Jylland, periglacial conditions prevailed with permafrost, solifluction, wind erosion and fluvial activity. Thus, the main ice advance led to a total rejuvenation of the ecosystem and when the final deglaciation began around 19 kyr BP it simultaneously set off a new vegetational succession and subsequent faunal regeneration. In this section the development of the terrestrial mammal fauna after LGM will be described and analysed

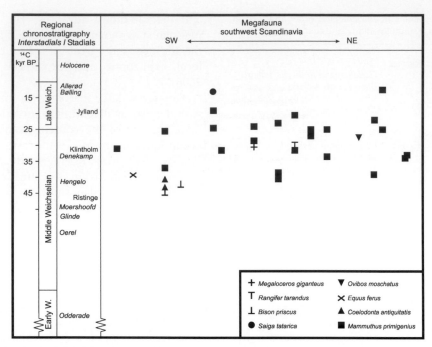

Fig. 25. Chronological and geographical distribution of dated Weichselian megafauna remains. Dates standing on the same vertical line share the same distance to the ice margin (see text for further explanation). (Revised after Aaris-Sørensen 2006).

especially with focus on changes in species diversity and composition.

Temporal distribution of species

The inferred late- and post-glacial time ranges given in the Systematic section are combined in a large species range chart (Fig. 26) including 53 terrestrial mammal species. The vast majority of the time ranges are based entirely on the subfossil record (solid lines) or on a combination of the subfossil record and knowledge about modern autecology and geographical distribution of the species in question (dotted lines). In three cases (*S. betulina*, *M. erminea* and *M. nivalis*) the range is mainly based on modern ecology and distribution and in four cases (*M. arvalis*, *A. agrarius*, *M. avellanarius* and *M. foina*) entirely so. These seven species have nevertheless been included in the chart and in the subsequent analyses as it seems desirable to make the database as complete as possible by including all native species still extant in the region. In order to decide whether this would affect the results, the diversity analyses were repeated without the seven species. As it can be seen by comparing Figures 27 and 28, the overall pattern is identical whether the seven species are included or not.

The species range chart reveals ten more or less comprehensive colonization events. The first took

place around 17 cal. kyr BP and was then followed by nine immigration waves around 14.5, 14.0, 13.5, 11.4, 11.0, 5.0, 2.5, 2.0 and finally 0.3 cal. kyr BP. The range pattern also shows that the first six colonization events were followed eventually by a number of extinctions, while all the species belonging to the last four are still extant members of the Danish mammalian fauna. The rate of extinction is high among the species belonging to the first three colonizations, but the highest number of still extant species, on the other hand, is seen among the 11.4 cal. kyr BP immigrants.

Changes in diversity and composition

In order to examine more closely how species diversity and composition have changed over time and to what extent these dynamics are linked to compensatory colonizations and extinctions, a set of analyses have been performed following the guidelines given by Brown *et al.* (2001).

Species richness is calculated by counting the number of terrestrial species recorded in each 1000-year interval and the long-term mean richness is the sum of richness counts divided by the number of intervals. Colonization is measured by counting the number of species present in one interval that were not present in the previous interval and, by contrast, extinction is determined by counting the number of species that

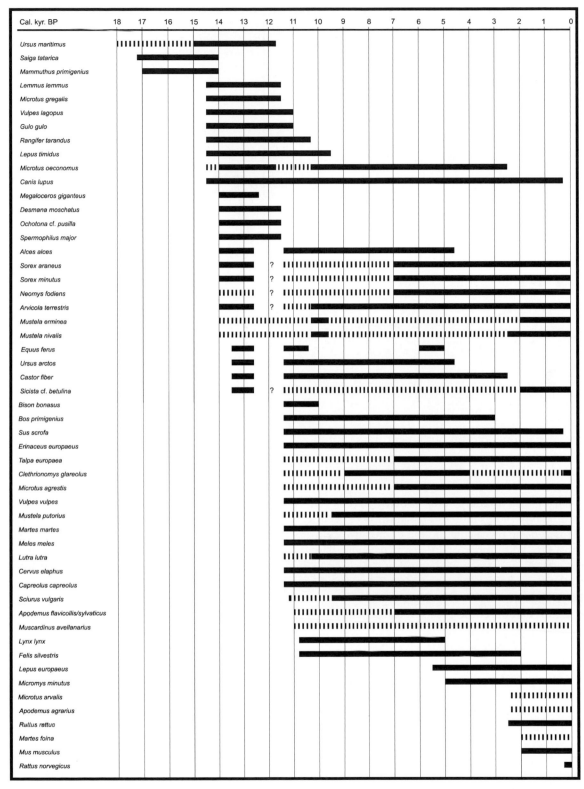

Fig. 26. Post-LGM species range chart. Solid chrono-lines are based on subfossil evidence while dotted lines are based partly or entirely on recent evidence. For details, see Systematic section.

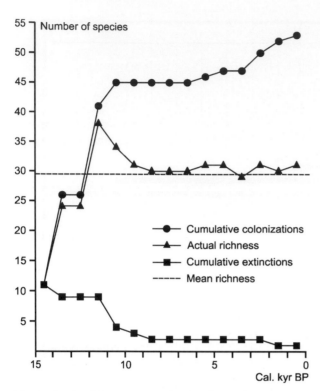

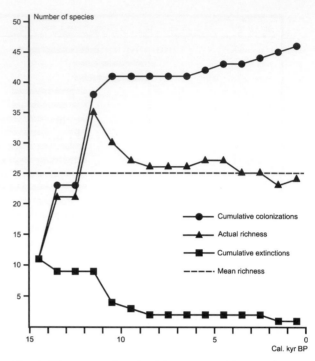

Fig. 27. Species richness, mean richness and cumulative colonizations and extinctions during the last 15 000 years.

Fig. 28. The pattern of species richness, mean richness and cumulative colonization and extinctions during the last 15 000 years when seven species with time ranges based mainly on recent evidence have been omitted.

were present in one interval but absent in the subsequent interval.

In order to assess the diversity curve further, it is also recommended by Brown *et al.* (2001) to quantify the changes in diversity that would have been observed if *only* colonization *or* extinction were occurring. This is done by calculating the cumulative colonizations, i.e. the number of species that would have accumulated if no extinctions had offset the observed colonizations and the cumulative extinctions, i.e. the number of species that would have remained if no colonizations had offset the observed extinctions.

Results

The data set makes a long-term study possible and Figure 27 shows the temporal variation in diversity within southern Scandinavia by means of 15 time slices representing values of richness for each 1000-year interval between 15 and 0 cal. kyr BP. A sharp increase in species richness can be seen during the first 4000 years going from an initial value of 11 to a final value of 38 species at the 12–11 kyr interval. This peak is followed by a less steep decrease over the next 2000 years reaching the value of 30 species at the 9–8 kyr interval. Since then, species richness has remained relatively constant through the last 9000 years with values a little higher

than the mean richness of 29 which is valid for the whole 15–0 kyr period.

To get an idea of the species composition lying behind the actual richness plotted on the diversity curve, the accumulated colonization and extinction curves have been added to Figure 27. The extinction curve illustrates the losses of taxa that would have occurred in a hypothetical closed system without the possibility of colonizations and the colonization curve the number of species that would have occurred if no extinctions had taken place. It is interesting to note that without recurring colonizations only two species would have been left as early as around 9–8 kyr (which in fact is the time of the highest actual richness) and that only one species would have survived into historic time (1–0 kyr). In the case of Denmark/ south Scandinavia, this species would have been the wolf, *C. lupus*, the only one to have been continuously present since the 14.5 kyr colonization event. It is likewise notable, in comparison with the mean richness of 29, to see the accumulated colonization curve end at 53 as the total number of species recorded in the region. The actual number of colonizations and extinctions are plotted in Figure 29 and the marked colonization events between 14.5 and 10.5 kyr with a break around the 13–12 kyr interval are easily recognized and so are the small-scale events between 5.0 and 0.3 kyr BP. The colonizations are to some extent

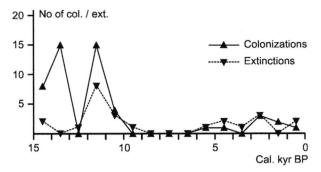

Fig. 29. Number of colonizations and extinctions occurring in each millennium throughout the last 15 000 years.

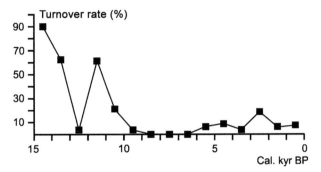

Fig. 30. Faunal turnover rate in each millennium throughout the last 15 000 years.

offset by subsequent extinction events between 12.5 and 9.5 kyr with a peak between 12.0 and 10.0 kyr and later from 5.5 to 2.0 and finally during Historic time (1–0 kyr interval).

Altogether, the accumulated as well as the actual colonization and extinction curves document large changes in species composition throughout the period. As a direct measure of the degree of change, the faunal turnover rate has been calculated for each 1000-year interval by expressing the sum of FADs and LADs as a percentage of the total number of species of the interval (Fig. 30). The plot shows very high turnover rates during the first 4000 years ranging between 90% and 60% followed by a decrease during the next 2000 years down to a 3000-year period with no change at all and a final 6000-year period with moderate turnover rates between ca. 4% and 20%.

Thus, the diversity of the mammalian fauna in Denmark/south Scandinavia after the LGM was characterized by a dramatic increase in richness between 15 and 11 cal. kyr BP followed by a decline towards a more moderate level between 11 and 9 kyr and then by maintaining a steady state the last 9000 years with a mean richness of ca. 30 species. The diversity was regulated by two opposing processes, colonizations and extinctions which also caused large changes in species composition.

Discussion

Climate and/or history

Ecologists and biogeographers intensively discuss whether the distribution and diversity patterns among present-day animal and plant communities are best explained by contemporary climate and environment or by historical conditions and changes in these (see Whittaker *et al.* 2001; Ricklefs 2004). In this context, history includes large-scale climatic events such as glaciations/deglaciations, subsequent changes in land/sea configuration, location of refugia and dispersal barriers. Besides being theoretically interesting, this discussion has direct relevance to modern nature conservation and biodiversity management especially in relation to future climate changes.

Recently, Svenning & Skov (2005, 2007) have shown that history is at least as important as current environment in controlling species composition and richness of European trees (except conifers) and more specific that LGM climate is a strong richness control for species with a restricted range, which appear still to be associated with their glacial refugia. Moreover, Araújo *et al.* (2008) support the view that temporal variation in condition of the climate over historic time can contribute to current species richness at least as much as contemporary climate. By testing species richness of all European amphibian and reptile species against climate of the LGM and the present-day climate variables (annual mean temperature and annual total precipitation sum), it appeared that climate stability between LGM and the present day is a better predictor of species richness than contemporary climate. The 0°C isotherm of the LGM turned out to delimit the distribution of narrow-ranged species, whereas the current 0°C isotherm limits the distribution of wide-ranging species. Hawkins & Porter (2003) have studied the relative influence of current and historical factors on mammal and bird diversity patterns in deglaciated North America. They found that factors acting in the present (temperature, precipitation, evapotranspiration, range in elevation and landcover types) appear to have the strongest influence but also that a historical signal (time after deglaciation) does explain 8–13% of the variance in species richness.

Danell *et al.* (1996) tested the hypothesis that current environmental features may correspond to the longitudinal variation in species richness of mammalian herbivores in the Holarctic boreal zone. They summed up, 'our study shows that present-day conditions, such as productivity and species–area relationships may play important roles in determining community composition of mammalian herbivores at

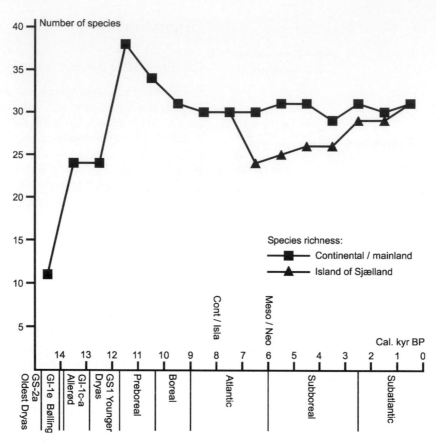

Fig. 31. Changes in species richness plotted against a climatic event chronology (Björck *et al.* 1998; Rasmussen *et al.* 2006). The transition between the continental and the island period (around 8.0 cal. kyr BP) and the transition between the Mesolithic and the Neolithic (around 6.0 cal. kyr BP) are also marked.

a large spatial scale'. And then they added, 'This is not to say that e.g. glacial history is unimportant. However, we have found it difficult to include such historical events in a quantitative and sensible way into our analyses'. Therefore, in the light of these discussions and for the benefit of future analyses of present-day diversity gradients, it may be useful and important to follow the *temporal* variation in diversity of the south Scandinavian mammal fauna after LGM in relation to climate-induced environmental changes.

In Figure 31 the richness curve has been connected with a climatic event chronology according to the isotopic Greenland ice-core record (Björck *et al.* 1998; Rasmussen *et al.* 2006). In trying to explain the forces driving the present-day global diversity gradient, the 'contemporary climate hypotheses' have focused on the correlation between richness and climate variables such as energy (temperature) and water availability. Hawkins *et al.* (2003) have reviewed the empirical literature in order to examine this relationship and found that measures of energy, water or water–energy balance explain spatial variations in richness better than other climate and non-climate variables in 82 of 85 cases. These cases include plants as well as

invertebrates and vertebrates. So, regional temperature and precipitation curves are given in Figures 32 and 33 to see whether the same correlation can be found with mammal richness over a time gradient of 15 000 years. The curves show the development of annual mean temperature and precipitation throughout the last 15 kyr for southern Scandinavia (54–58°N, 8–14°E). The climate curves are the results of Late Quaternary Atmosphere–Slab Ocean Simulations coupled to General Circulation Model Simulations kindly provided by Alan Haywood and Paul Valdes (Earth System Modelling Results from BRIDGE, The Bristol Research Initiative for the Dynamic Global Environment). In these simulations, the mean annual temperatures and annual precipitation sum for the LGM to the pre-industrial era were calculated using a suite of 23 climate simulations (pre-industrial, LGM and Younger Dryas plus one simulation each 1000 years of time from the LGM to 1000 years BP) derived from a General Circulation Model (GCM). The GCM used is the HadAM3 version of the UK Meteorological Office's Unified Model (Wood *et al.* 1999) coupled

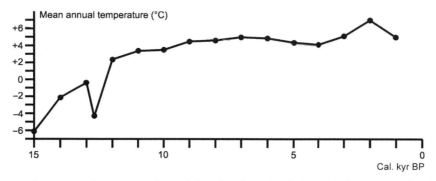

Fig. 32. The development of mean annual temperature in south Scandinavia during the last 15 000 years. (Based on BRIDGE Earth System Modelling Results, see text for further explanation).

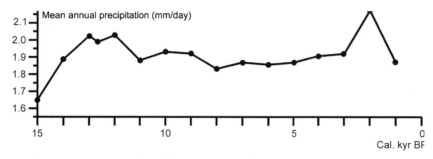

Fig. 33. The development of precipitation, expressed as daily mean, in south Scandinavia during the last 15 000 years. (Based on BRIDGE Earth System Modelling Results, see text for further explanation).

to the slab-ocean model HadSM3, which is a standard Q-flux ocean model (Hewitt *et al.* 2001).

Finally, two marked events have been plotted on the time-scale of Figure 31 that is the Early Atlantic transgressions around 8 kyr BP ending the Continental Period and leading to the formation of the Danish Islands (see section on Islands) and the Mesolithic/-Neolithic transition around 6 kyr BP with the introduction of agriculture and husbandry leading to progressive deforestation.

The vegetational development during the highly changeable Late Glacial and Early Holocene is documented in Figure 34. The figure presents a pollen percentage diagram for trees, shrubs and herbs from a small kettle hole located in the southern part of Jylland at Slotseng near Vojens (Mortensen 2007). Although Slotseng is a small-scale site which primarily reflects the local environmental changes, most of the variations in the pollen assemblage can, nevertheless, be ascribed to known bio- or climatostratigraphical phases in north-western Europe/southern Scandinavia. Besides, the Slotseng stratigraphy is the only well-dated locality known to contain an undisturbed stratigraphy from the first Late Glacial warming to the Early Holocene, ca. 15.5–10.8 cal. kyr BP (Mortensen 2007). Added to this diagram is the loss-on-ignition (LOI) curve as an indirect measure of the biological productivity in the former lake basin.

The peak, 15–11 cal. kyr BP (habitat migration, refugia, non-analogue faunas, individualistic response)

The peak in mammal richness is built up from 15 to 11 kyr BP by four colonizations at 14.5, 14.0, 13.5 and 11.4 kyr BP. The dramatic increase in richness from 15 to 11 kyr BP follows a marked increase in temperature and precipitation. According to the BRIDGE Earth System Modelling Results (see above), the annual mean temperature in south Scandinavia rises from −6.1 to +3.4°C involving a rise in summer temperature (mean JJA) from +10.8 to +16.0°C and an even more pronounced rise in winter temperature (mean DJF) from −21.8 to −7.3°C. Annual mean precipitation is estimated by the BRIDGE model to increase from 604 to 687 mm.

There is a break in the increase in richness during the 13–12 kyr interval. This corresponds to the well-known Younger Dryas cooling (Greenland Stadial 1) which according to the Greenland ice core chronology had a duration of 1186 years with the onset at 12.9 kyr b2k (before AD 2000) and with a cold maximum a few centuries later (Rasmussen *et al.* 2006). No colonizations took place in this interval, but if we look at the range of dates of several large mammals (*C. fiber, U. arctos, A. alces, M. giganteus* and *E. ferus*), it suggests that the Younger Dryas cooling led to a

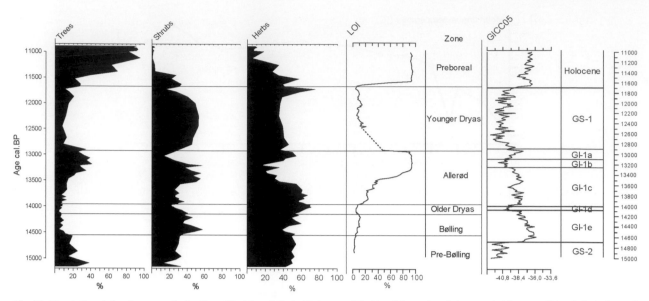

Fig. 34. Vegetational development during Late Glacial and Early Holocene (15–11 cal. kyr BP) as it is recorded in a small kettle hole located in the southern part of Jylland at Slotseng near Vojens. LOI curve (loss-on-ignition) can be seen as a proxy for biological productivity in the former lake basin. Local pollen zones are added as well as the Greenland ice core chronology according to Björck *et al.* (1998) and Rasmussen *et al.* (2006). (After Mortensen 2007).

local extinction of these and possible other species (Fig. 26). This Younger Dryas Induced Pause lasted from ca. 12.6 kyr BP until a re-immigration took place at the beginning of Holocene around 11.4 kyr BP. As the species (with the exception of *Megaloceros*) all re-appear in the subsequent 12–11 kyr interval, they are not registered here as actual extinctions.

If the LOI values in the Slotseng basin (Fig. 34) are regarded as a proxy record of the changes in biological productivity in the area, the general rise in energy input can be followed through the 15–11 kyr period. The 15–14 kyr interval is characterized by very low but gradually increasing LOI values followed by a rapid and sharp increase in the 14–13 kyr interval reaching a plateau with stable values around 13.3 kyr BP. This sequence corresponds to the first step of the rise in mammal species richness. A sharp decrease to low LOI values again can be seen between 12.9 and 11.7 kyr BP corresponding to the Younger Dryas cooling and this low is once more followed by a sharp rise reaching a new plateau in Early Preboreal. This rise corresponds to the second step in the rising curve of mammal species richness.

The contemporary shift in vegetation around the Slotseng basin has been recorded through pollen- and macrofossil studies (Mortensen 2007). During the early phases of the 15–14 kyr interval, the local upland is sparsely vegetated and strongly influenced by periglacial conditions favouring a stress-tolerant pioneer vegetation of *Salix–Betula nana–Dryas octopetale*. Later changes are seen towards a more permanent pioneer vegetation with arctic/subarctic

calcareous grassland or *Dryas* heath with single elements of steppe vegetation. During the first half of the 14–13 kyr interval, there are changes towards taller vegetation dominated by *Salix* and grasses. This open *Salix*/grass community exists until the first tree birch appears in the beginning of the second half of the interval. Tree birch probably grew in the moist areas around the lake and *B. nana* on the drier and open areas. The cold Younger Dryas period is dominated again by *B. nana* and *Salix* which is rapidly replaced by shrub vegetation dominated by *Juniperus communis* and *Empetrum* in the earliest Preboreal. After approximately 200 years the tree birch expanded into the area again and an open *Betula pubescens–Populus tremula* forest was created. The local presence of *Pinus* is documented by trunks and branches found in the Slotseng sediments and dated to ca. 10.3 cal. kyr BP.

As previously noticed (Aaris-Sørensen 1992) it appears that the dispersal of mammalian species into the former glaciated southern Scandinavia occurred almost instantaneously as soon as favourable habitats had been formed. So, the expansion into the deglaciated areas could not primarily be dependent on the migration abilities of the species and/or the distance to possible refugia in southern Europe. In the Systematic section, many examples were given of species being present in areas adjacent to southern Scandinavia well in advance of the actual colonization. The mammalian fauna in the Neuwied Basin, Central Rhineland, for instance, has been studied intensively (e.g. Turner 1990; Street &

Baales 1999) and reveals the presence of e.g. *S. minutus, T. europuea, L. timidus, O. pusilla, C. fiber, C. glareolus, A. terrestris, M. oeconomus, C. lupus, V. vulpes, M. nivalis, M. meles, E. ferus, S. scrofa, C. elaphus, A. alces, R. tarandus, C. capreolus* and *B. primigenius* in the time *after* the south Scandinavian deglaciation but 2–3500 years *before* they finally immigrated into Denmark and southern Sweden. These mammals are generally highly mobile and good dispersers, the distance was not greater than that covered during seasonal migration by some of the larger species and no severe geographical barriers existed. So, the mammals waited for a balance between climate and vegetational conditions to be established and the rate of this 'habitat migration' governed the expansion into the new areas.

The role of potential temperate refugia located in southern Europe (Iberia, Italy and the Balkans) in the shaping of modern distribution and genetic diversity of species has been given much attention (Taberlet *et al.* 1998; Hewitt 2000). Recent studies, however, indicate a much wider distribution of trees in central and eastern Europe during LGM (Willis *et al.* 2000) and the presence of cryptic northern refugia for trees and mammals in areas of sheltered topography has been suggested by Stewart & Lister (2001). This could explain the rather early occurrence after LGM of many mammalian species in northern Europe as the presence of wide-spread refugia offers an alternative to the rapid long-distance colonization from the three southern peninsular refugia. Besides, the spread from multiple refugia could have led to complex and varied genetic interactions among populations as suggested by Stewart & Lister (2001, box 1). The south Scandinavian region is important for understanding the re-colonization of northern Europe and the Fennoscandia and the Late Glacial/Early Holocene bone remains presented here should play an important role in future, ancient DNA-based, phylogeographical studies.

Studies of primary succession in recently deglaciated terrain on glacier forelands very often show a peak in richness early in succession followed by a decline (Matthews 1992). Early peaks seem to be characteristic of open areas with favourable conditions attracting a relatively large number of species and also to reflect an overlap in the distribution of pioneers and later colonizators. The subsequent decline is believed to be caused by increasing competition as the vegetation cover approaches 100%. Although these studies are confined to plant communities and the time ranges are short (200 years maximum), it is nevertheless interesting to recognize the same shape in the long-term mammalian succession after LGM. Likewise, it seems obvious

that the large number of mammalian species forming the peak during the 15–11 kyr period is an overlap of early 'pioneers' in the form of species today characteristic of tundra and steppe environments (e.g. *L. lemmus, L. timidus, R. tarandus, D. moschata, O. pusilla* and *S. major*) with later colonizators today characteristic of taiga and temperate forest (e.g. *E. europaeus, C. fiber, S. scrofa, C. elaphus, A. alces, C. capreolus* and *B. primigenius*). A mixture like this containing several species which today are considered ecologically incompatible has been referred to as non-analogue or disharmonious assemblages (Graham & Semken 1987; Graham *et al.* 1996). In a case study (Aaris-Sørensen 1995) comparing the small mammal fauna revealed from a Late Weichselian freshwater bed at Nørre Lyngby in Northern Jylland with the modern fauna in the same region, it has been shown that the fossil fauna has the highest and almost equal proportions of boreal forest and steppe elements followed by a relatively high proportion of tundra elements. The modern fauna, on the other hand, has a preponderance of species with deciduous woodland affinities followed by boreal elements. If one looks for a present area of sympatry for the fossil species from Nørre Lyngby, it can be found north of the Caspian Sea from the river Volga in the west to the southern and western slopes of the Urals. If, however, the fossil small mammal fauna is supplemented with the fossil large mammals known from the area (and from other Danish Late Weichselian sites) then the fauna has no modern analogue. The fossil fauna contains species that are today allopatric; so, the area of sympatry is 'disharmonious' with modern distributions (Graham & Semken 1987). The peak in mammal richness is composed of a non-analogue, disharmonious fauna which makes the Early Preboreal the most diverse period in the history of Danish mammals.

Related to the recognition and description of non-analogue fossil assemblages is the attempts to understand how terrestrial mammal communities respond to environmental change during Late Pleistocene. Two models have been advanced. One (the Clementsian or deterministic) sees communities as tightly organized, co-evolved and long-lived groups of species where all species will respond to environmental changes in the same way. The other (the Gleasonian or individualistic) assumes that each species will respond in accordance with its individual tolerance limits and therefore disperse diachronically in different directions and at various rates (Graham *et al.* 1996). The models have been tested against a large database (FAUNMAP) containing 2945 late Quaternary mammal localities in the USA. These documented that for the last 20 000 years mammals responded in a

Gleasonian manner to environmental changes creating non-analogue faunas (Graham *et al.* 1996). The composition of the Danish Late Glacial/Early Holocene mammal fauna as described above can only likewise be explained as the result of individualistic responses by species dispersing with varying rates, at different times and in different directions.

Decline and steady state, 11–9 and 9–0 cal. kyr BP (neolitization, deforestation, increasing human impact)

The peak in richness is followed by a decline between 11 and 9 kyr BP as a consequence of the progressive extinction of species belonging to the early 14.5 and 14.0 kyr BP colonizations, which were adapted to the young ecosystem with open environments. Combined with a marked new colonization event at 11.4 kyr BP, this produced a faunal turnover rate of 60% and a decline in actual richness from 38 to 31 species; only two of the species from the 14.5 colonization were left. During this decline the temperature and precipitation were still rising through the Preboreal and Boreal periods and the vegetation changed from an open birch–pine forest through a more dense hazel dominated woodland towards a still denser mixed deciduous forest (Fig. 3). The faunal turnover represented the exit of all arctic, subarctic and steppe species and the arrival of species more adapted to forest environments.

From 9 kyr BP and onward the richness curve describes a steady-state situation with an actual richness around 30. Throughout the first three millennia no colonizations and no extinctions occurred but during the last six millennia the faunal turnover rate varies between 4% and 20% constantly changing the species composition. According to the BRIDGE Earth System Modelling Results (see previous section on Climate and/or history), the temperature and precipitation has been almost constant during these 9000 years which covers the Atlantic, Subboreal and Subatlantic periods, although with a rise around 2.0 kyr BP (known as the Roman Warm Period).

The faunal changes after 6.0 kyr BP coincides with the progressive deforestation caused by the introduction of agriculture and husbandry in southern Scandinavia (Fig. 3).

This man-made habitat fragmentation had an effect on both the vulnerable, and already impoverished, island fauna (se below) and on the continental mainland fauna. At the same time a growing human population began to regard most wild mammals as competitors for natural resources and a threat to their crops and livestocks. Together, this led to extinctions, but simultaneously the new open landscape made colonizations possible for other species.

As seen in the species range chart (Fig. 26) *L. lynx, U. arctos, A. alces* and *B. primigenius* became extinct in Jylland between 6.0 and 3.0 kyr BP; extinctions common to both the islands and the mainland took place during the 3–2 kyr interval of *M. oeconomus, C. fiber* and *F. silvestris* and in historical times of *C. lupus* and *S. scrofa*. The final extinctions of the wolf and the wild boar in the AD 1700s are known to be the result of a severe and persistent hunting pressure and in the case of the wolf even a result of an official extermination campaign. A similar campaign was launched in 1799 against the red deer as it was believed to damage forests and farm land severely (Weismann 1931). Fortunately, it only became a limited success as the mainland population in Jylland managed to survive. It is hard to know how far back actual persecution of a certain species can be recognized. In the case of the beaver (*C. fiber*) it is, nevertheless, interesting to see that the final extinction around 2500 BP coincides with an intensified agricultural exploitation of the landscape in the early Iron Age (Rasmussen *et al.* 2007). This might very well have caused a conflict of interests concerning the luxuriant and economically important meadows along rivers and streams.

In terms of sheer numbers the extinctions are counterbalanced by successive colonizations of *L. europaeus, M. minutus, M. arvalis, A. agrarius, R. rattus, M. foina, M. musculus* and *R. norvegicus* (Fig. 26). The hare was attracted by the open landscape itself, the harvest mouse, striped field mouse, common vole and beech marten by the man-made cultural landscape and the house mouse and the rats are directly introduced by man and closely associated with human habitation and activities.

Islands (impoverishment, size reduction)

It should be remembered that the analyses of species diversity and composition (Figs 26–30) disregard the fact that Denmark was transformed into a group of islands and peninsulas around 8 kyr BP (Fig. 35). Strictly speaking, the curves beyond 8.0 kyr BP only represent the Jylland peninsula which has been part of the North European plains all through the Weichselian and the Holocene. If we look separately at the islands, some of the larger ones reveal major changes in faunal composition as early as in the 8–7 kyr interval.

The results of faunal analyses conducted by the author in the 1970s on four Mesolithic sites (Kongemose and Ertebølle cultures) near Vedbæk in northern Sjælland indicated a local extinction of several mammalian species in the early Atlantic period around 8–7 kyr BP. To see whether this pattern was valid for the entire island of Sjælland, fauna lists from 23 other Atlantic settlements and stray finds from non-archaeological contexts were included. A total of some 50 000 mammal bones then showed the same pattern with the total absence of remains of *U. arctos*, *M. putorius*, *M. meles*, *L. lynx*, *A. alces* and *B. primigenius* (Aaris-Sørensen 1980; Tables 1, 2).

These six species are well known as stray finds from Preboreal and Boreal deposits and from Maglemosian sites from all over the area which later became the island of Sjælland. Furthermore, after the creation of the Danish islands, the six species continued to be common on the Jylland peninsula another 3–8000 years with two of them still extant (*M. putorius* and *M. meles*). Today after additional 30 years excavating and collecting the pattern still holds. Now, the ZMUC collection and files contain 161 Late Mesolithic and Neolithic sites from the island of Sjælland with mammalian remains but without any of the six species. In Jylland, on the other hand, 49 contemporary sites all yield one or more of the six species. Four of the species are represented by large numbers of dated specimens, but it should be mentioned that the lynx and the polecat are represented by only 16 and 20 dated specimens respectively. Future discoveries may tell whether the absence of evidence of the two carnivores on Sjælland is the matter of chance or reflects the true distribution in time and space. A review of the few single specimens of the six species that actually have been found in archaeological contexts younger than the postulated time of extinction has been given by Aaris-Sørensen (1998, pp. 178–189). These are typically single ornamented bone tools or grave goods as tooth pendants none of which can be considered as true indicators of the local fauna.

A possible early extinction of the two ungulates *A. alces* and *B. primigenius* in the island of Sjælland was suggested for the first time by Degerbøl (1933, p. 378). Later Degerbøl (1964, p. 81, 1970, pp. 131–135) considered the Early Atlantic extinction a fact and considering the causes for the extinction he pointed out that three main factors are to be taken into account: (1) the change in vegetation from the open Boreal forest to the dense Atlantic climax forest; (2) the impossibility of immigration for new individuals from the continent; and (3) hunting pressure by humans. Aaris-Sørensen (1980), dealing with a more broad-spectred extinction including six species of carnivores and ungulates, was more reluctant to point out a common combination of biotic and abiotic factors leading to the extinctions. He regarded the formation of the island itself and the resulting isolation as the crucial factor. It is a long known fact that continental islands exhibit an impoverished fauna compared with the nearby mainland (MacArthur & Wilson 1967; Udvardy 1969). Besides, island faunas show a general high vulnerability to changes. With water barriers preventing or reducing new immigrations, otherwise harmless population fluctuations caused by climatic or biotic changes (as e.g. changes in forest composition or human impact as suggested by Degerbøl) might become a disaster to an island population and therefore lead to extinction.

Another well-documented characteristic among island mammals is the tendency towards either dwarfism or gigantism (e.g. Sondaar 1977; Lister 1989; Vartanyan *et al.* 1993). Van Valen (1973) has named it 'the island rule' by which island rodents (and perhaps marsupials) tend towards larger size, while carnivores, lagomorphs and artiodactyls tend towards smaller size, and insectivores show neither tendency.

Jensen (1993) has measured 1505 fossil bones and teeth of Danis roe deer (*C. capreolus*). The fossil bones were divided into three groups, Boreal continental, Atlantic mainland and Atlantic island (the islands of Sjælland, Fyn and Lolland) sample, and these groups were compared both with one another and with measurements taken on 54 recent skeletons. The investigations showed that the animals from the Boreal continental as well as the two Atlantic samples were much larger than the recent roe deer, but also that the insular Atlantic animals were significantly smaller than both the Boreal continental *and* the contemporaries from the mainland. The smaller size of the insular roe deer could be seen as an adaption to smaller areas, higher population densities and less food (MacArthur & Wilson 1967; Wasserzug *et al.* 1979). In this way a selection for smaller size can postpone or even, as in the case of the roe deer on the Danish islands, prevent extinction.

The large amount of data from Sjælland has made it possible to plot a separate richness curve for this island covering the last 8000 years (Fig. 31). Attempts in the same way to analyse the island of Fyn (including the South Fyn Archipelago only separated from the main island by a very narrow sound of 175–200 m) and the island of Bornholm, which was isolated much earlier than the other Danish islands, have been made previously by Aaris-Sørensen (1985, 1998). Although the data are too few to support any clear conclusions it seems that the island of Fyn experienced an extinction of the three large carnivores at the same time as Sjælland, whereas the aurochs, elk and polecat survived another 2–3000 years. The final isolation of Bornholm took place around the transition Preboreal/Boreal a little more than 10 000 years ago (Björck 1995). Today the island has 12 terrestrial mammalian species of which only six are known from the fossil record.

Among the 13 species known as fossils only *R. taran-dus*, *A. alces* and *S. scrofa* are well documented and it can be seen that the extinction of the reindeer coincided with the extinction of the entire south Scandinavian population while Bornholm was still connected to Pommern, whereas the elk disappeared from the island shortly after its isolation, but the wild boar survived at least another 6000 years until the end of the Atlantic period.

The drop in richness on the island of Sjælland during the 8–7 kyr interval is followed by a gradual climb back to the mainland curve (Fig. 31), which is reached again during historical/recent times. This is due to the facts that two of the extinct species, the badger and the polecat, reappear on the island during the 2–1 kyr and 1–0 kyr interval, respectively, and that the mainland eventually suffers the same extinctions of *Bos*, *Alces*, *Ursus* and *Lynx* at the same time as Sjælland is affected by the same new colonizations as the mainland. The re-appearance of the polecat on the island of Sjælland (and Fyn) is supposed to be the results of animals escaping fur farms during the 1930s (Jensen & Jensen 1972). It has been believed that the extant badgers of Sjælland and Fyn alone were the results of reintroductions made by humans during the 1800s on the islands of Fyn and Lolland–Falster. Recently, however, two badgers from Søborg and Æbelholt in northern Sjælland have been radiocarbon dated to 1305 ± 50 ^{14}C yr BP (LuS 7372) and 995 ± 50 ^{14}C yr BP (LuS 7373) corresponding to ca. AD 700 and AD 1000 respectively. They were selected among a small sample of eight badgers found in archaeological context and believed to date between Iron Age and Medieval times. So, the re-appearance of the badger on the island of Sjælland apparently took place more than a 1000 years earlier than believed, either by a natural re-immigration or a man-made reintroduction.

The present species richness on the Danish islands has recently been described using data from the Danish Mammal Atlas Project (Christiansen & Jensen 2007). In all, 47 islands were included, spanning in size from 1 to 100 000 ha and in richness from 1 to 28 species (the islands of Fyn and Sjælland not included). It could be concluded, in accordance with MacArthur & Wilson (1967), that the most important factor in determining present species richness is the size of the island. The larger the size, the larger the number of species found. Also important, but less so, was the distance to the nearest mainland or large island. The shorter the distance, the larger is the number of species. Apart from the three largest islands Sjælland, Fyn and Bornholm, it will probably never be possible to unravel the history underlying the present species composition on the Danish islands. The

Fig. 35. South Scandinavia before and after the Early Atlantic transgressions. Top: the continental period around 9.0–8.5 cal. kyr BP. Bottom: the island period around 6.5 cal. kyr BP. (Drawing: Knud Rosenlund).

modest fossil and historical data, however, indicate a very dynamic history with repeated extinctions and re-colonizations events on every island. Common to the most abundant island species today is the fact that they are habitat generalists. According to Christiansen & Jensen (2007), this is a consequence of the Danish islands being highly diverse in their natural environment and human impact.

The marine fauna, ca. 115–0 kyr BP

The subfossil record of marine mammals in Denmark comprises 15 whale species, five seal species and the polar bear (the latter is deliberately treated as part of the terrestrial group as well).

The ZMUC files contain around 360 whale and 260 seal specimens of which ca. 55% of the whales and ca. 90% of the seals have been found in an archaeological context. Represented in the largest numbers among the archaeological findings are the remains of smaller toothed whales, especially the harbour porpoise, *P. phocoena*, and among the seals the grey seal, *H. grypus*.

The remaining ca. 45% of the whales and 10% of the seals have been found in a geological context either *in situ* or redeposited. These remains mainly represent large baleen whales or larger toothed whales, and among the seals especially the walrus, *O. rosmarus*, and the ringed seal, *P. hispida*. The single find of polar bear also belongs to this group. The bones have been fished or dredged from the bottom of the sea or excavated in raised marine sediments, including old beach ridges, or found redeposited in glacial and glaciofluviale deposits. Except for a hind part of a bowhead whale, *B. mysticetus*, a front part of a fin whale, *B. physalus*, and three complete skeletons of ringed seals, all excavated in raised marine sediments in northern

MARINE STAGES & ENVIRONMENTS (mollusc faunas)*	OLDER YOLDIA SEA			YOUNGER YOLDIA SEA	OLDER TAPES SEA	YOUNGER TAPES SEA
Climate	boreal / boreal-arctic	arctic-subarctic		arctic-subarctic boreal	boreal-lucitanian	
Stage	E. WEICH.	MIDDLE WEICHSELIAN		L. WEICH.	HOLOCENE	
Range of dates	>40.0	33-21	21-11.7	11.7-5.7	5.7-0	
Ursus maritimus Polar bear	1N	1N	2			
Odobenus rosmarus Walrus		3+7				
Phoca groenlandica Harp seal		4N	5S	6	6	
Phoca hispida Ringed seal	7	4N	7+8	7	7	
Phoca vitulina Common seal				7	7	
Halichoerus grypus Grey seal				7	7	
Lagenorhynchus albirostris White-beaked dolphin				9	7	
Delphinus delphis Common dolphin				7	7	
Tursiops truncatus Bottlenose dolphin				7	9	
Stenella sp. (Striped) dolphin				10		
Orcinus orca Killer whale			11S	9	9	
Phocoena phocoena Harbour porpoise			5S	7	7	
Delphinapt. l. / Monodon m. Beluga whale / Narwhale	9	9	9	12		
Hyperodon ampullatus Northern bottlenose whale			11S	7		
Physeter macrocephalus Sperm whale				7	9	
Balaenoptera acutorostrata Common minke whale		9		9	9	
Balaenoptera physalus Fin whale					9	
Balaenoptera cf. *musculus* Blue whale					9	
Megaptera novaeanglia Humback whale			9		9	
Eubalaena glacialis Atlantic right whale			9	9		
Balaena mysticetus Greenland right whale			9		9	

Fig. 36. Range of dates of marine mammals. Sources; 1 = Lauritzen *et al.* (1996); 2 = Aaris-Sørensen & Petersen (1984); 3 = Møhl (1985); Aaris-Sørensen (1998); 4 = Hufthammer (2001); 5 = Fredén (1984); 6 = Bennike *et al.* (2008). 7 = Unpublished ZMUC files; 8 = Lagerlund & Houmark-Nielsen (1993); 9 = Aaris-Sørensen *et al.* (in press); 10 = Trolle-Lassen (1985); 11 = Lepiksaar (1966); 12 = Winge (1899). N, Norwegian west coast. S, Swedish west coast. *Climatic affinities after Petersen (2004).

Jylland, the rest of the marine mammals are only represented by one or a few bone elements.

As in the history of the terrestrial mammals, the distribution of the marine mammals in time and space is determined by the dynamic changes in the land/sea/ice configurations. Seen in the right geological perspective, the present Late Holocene geography with its extensive Inner Danish Waters can only be seen as an unusual situation. The normal situation throughout the last glacial and late glacial period was characterized by a sea restricted to northern Denmark (see Fig. 2C, H). A narrow, funnel-shaped Kattegat embayment normally reached as far south as northern Sjælland, connected further south through a narrow strait with a Baltic Ice Lake and further north through the Skagerrak and the Norwegian Channel with the North Sea. The palaeo-Kattegat–Skagerrak also covered the present-day Vendsyssel. This overall geography was maintained throughout the period, although with some modifications first of all during the relatively short-lived glacial advances (see Fig. 2B, F). Based on a wide range of sedimentological and biostratigraphical analyses different marine stages with specific environmental conditions have been recognized (Fig. 36).

Figure 36 summarizes the primary evidence given in the Systematic section concerning the chronological distribution of the marine mammals in relation to the marine stages and changes in environment. Some of the records known from the Norwegian and Swedish west coasts have been included as far as they extend the time range given by the Danish record alone. Current evidence has been added as well for each species in the Systematic section in order to infer a more correct time range. This suggests that besides *U. maritimus*, *P. hispida* and *D. leucas* also *O. rosmarus*, *P. groenlandica*, *L. albirostris*, *O. orca*, *P. macrocephalus*, *B. acutorostrata*, *E. glacialis* and *B. mysticetus* have been present, at least occasionally, during all the marine stages of the Weichselian.

A closer look at the geographical spread of the dated remains shows that the oldest dates are confined to northern Jylland (and the west coast of Norway and Sweden) and to cold-adapted species such as *U. maritimus*, *O. rosmarus*, *P. hispida*, *P. groenlandica* and *D. leucas*. This is in agreement with the geographical distribution of the arctic–subarctic palaeo-Kattegat–Skagerrak as described above. The youngest dates, on the other hand, are found to be much more spread and much further south, and they include warm-adapted species such as *D. delphis*, *T. truncatus* and *Stenella* sp. This again is in accordance with the Mid–Late Holocene transgressions creating the Inner Danish Waters.

Polar bear

The single Danish record of polar bear, which was found in northern Jylland, represents, together with seven other specimens known from the south-western coast of Norway and the west coast of Sweden, the southernmost and youngest occurrence of the species in Europe. They date between ca. 15.5 and 12.0 cal. kyr BP corresponding to the arctic–subarctic Younger Yoldia Sea and to the last existence of drift ice, pack ice and icebergs in the Kattegat–Skagerrak area.

Walrus

Four of the five walrus specimens known from Denmark have been radiocarbon dated and the dates fall between ca. 31.0 and 23.5 ^{14}C kyr BP. They all show clear signs of having been transported by water and/or ice; three of them were dredged from the North Sea bottom and two found in gravel pits in Northern Jylland. They must have been brought to their finding place by later ice advances coming from north and/or north-east. Exact positions were given by the fishermen in two cases, namely 40 nautical miles NW of Lyngvig Fyr and 6–7 nautical miles NW of Hanstholm; so, all specimens have been finally deposited within the range of the main ice advance reaching the Main Stationary Line around 22 cal. kyr BP (Fig. 2F). The oldest walrus dates to around 31 ^{14}C kyr BP and must, according to the palaeogeographical reconstructions given by Houmark-Nielsen *et al.* (2005), have lived in the Older Yoldia Sea somewhere in the Skagerrak–Kattegat area. However, the three other dates lie between ca. 27 and 24 ^{14}C kyr BP and these walruses should therefore, according to the same palaeogeographical reconstruction, have lived in an ice-dammed lake, the Kattegat Ice Lake. From a biological point of view this is not likely, for which reason an adjustment of the palaeogeography by redating the sediments and fossils involved, including the walruses, is necessary.

True seals

Four species of true seals are known from the subfossil record, *P. groenlandica*, *P. hispida*, *P. vitulina* and *H. grypus*. The two last-mentioned are only known from the Holocene, whereas the two first mentioned are known from the Weichselian as well. While the harbour seal, *P. vitulina*, is by far the most common and most numerous seal in the present Danish waters and the grey seal, *H. grypus*, is rare, quite the opposite situation seems to have prevailed in the past. More than half of all the seal remains known from the Holocene belong to the grey seal, whereas only ca. 8% belongs to the harbour seal. The few remains of the harbour seal point at a continuous but sporadic occurrence in the Danish waters during the last 8000 years but give no indication of when it became a common species along our coastal areas. We know, however, that the population was considered a severe threat (together with the grey seal) to fishing during the 1800s and in order to decimate the stock a government-financed bounty was introduced in 1889 lasting until 1927 (Joensen *et al.* 1976). The intensive hunting also affected the grey seal which was almost wiped out at the beginning of the 1900s (Tougaard 2007).

Figure 36 shows that the chronological distribution of the arctic–subarctic ringed seal, *P. hispida* and the harp seal, *P. groenlandica* are not, as might be expected, confined to the arctic–subarctic Yoldia Sea of the Weichselian. Both species are also known from the Holocene by ca. 75 specimens altogether.

The ringed seal has been found on 25 different archaeological sites covering the last 8000 years. Except for a few finds from northern Norway, all other subfossil ringed seals from Europe have been found in and around the Baltic Sea (Sommer & Benecke 2003). As its presence in the Late Glacial Kattegat–Skagerrak is well documented (Systematic section this study; Lepiksaar 1966) it is presumed that the ringed seal entered the Baltic during the deglaciation of south-central Sweden around 10.3–10.2 ^{14}C kyr BP (ca. 12.0 cal. kyr BP). At this time, the Baltic Ice Lake was drained into the Younger Yoldia Sea (Kattegat–Skagerrak) as the retreating ice could no longer withhold the dammed-up water masses (Björck 1995). The connection between the Baltic basin and the North Sea across south-central Sweden lasted until ca. 9.5 ^{14}C kyr BP (ca. 11.0 cal. kyr BP) when it was broken by continuing glacio-isostatic land uplift and the creation of the Ancylus Lake stage. So far, the earliest radiocarbon date of a ringed seal in the northern Baltic comes from Nurmo on the Finnish west coast with an age of 9.5 ^{14}C kyr BP (Ukkonen 2002). So, the ringed seal was present in the northern Baltic from the very beginning of the Ancylus Lake stage (at the latest) which support the idea of an imigration across south-central Sweden. However, Ukkonen (2002) is right to remark that an even earlier immigration during the Baltic Ice Lake stage through the outlet in Øresund, or along ice-dammed lakes and rivers, cannot be entirely excluded. Anyhow, in relation to the Danish seal record it is important to realize that a Baltic ringed seal population has existed during the last 11 000 years and that the few Danish remains probably represent occasional expansions of this population or the visit of single stragglers.

The presence of the arctic harp seal in Danish waters and the Baltic during the Holocene climatic optimum has been much debated (e.g. Degerbøl 1933; Lepiksaar 1964; Møhl 1971a; Fredén 1984; Lougas 1998; Storå 2001; Ukkonen 2002; Storå & Lougas 2005). The main question discussed has been whether the remains represent a former breeding population in the Baltic or yearly or occasional migrations from the arctic. Storå & Ericson (2004) have recently looked at the problem once more by analysing harp seal bones from 25 archaeological sites from the Baltic and Danish Waters. Comparisons of bone measurements and epiphyseal fusion data of archaeological and recent harp seals indicate that young, most likely newborn, harp seals were common in the Baltic in the Subboreal. Besides, subadult harp seals of all age classes occurred, indicating that harp seals were present in the Baltic throughout the year.

Out of the 49 Danish records 24 specimens were recently selected for radiocarbon dating. The results show (Bennike *et al.* 2008) that the oldest date lie around 6.0 cal. kyr BP, indicating that the harp seal arrived several millennia after full marine conditions were established in Danish waters. The earliest date of harp seal recorded from the Baltic Sea is slightly older around 6.4–6.2 cal. kyr BP (Ukkonen 2002). Furthermore, it can be seen that harp seals were common in Danish waters during two periods one from around 6.0 to 5.8 cal. kyr BP and another from around 5.0 to 4.4 cal. kyr BP. Archaeological and geological records seem to suggest the existence of two contemporary periods with enhanced salinity in the Danish waters and the Baltic. Therefore, it is concluded that the temporal distribution of harp seal dates reflects environmental changes rather than exploitation patterns. Taking into consideration the evidence of a local Baltic breeding population as presented by Storå & Ericson (2004), it is suggested by Bennike *et al.* (2008) that hunger migrations by the Jan Mayen population led to the establishment of a harp seal population in the Baltic Sea in the Late Atlantic around 6.4 cal. kyr BP. These migrations are well known from historical and modern times and seem to follow a sharp decline of the Barent Sea capelin (*Mallotus villosus*) stock (Gjøsæter *et al.* 2003). Huge numbers of harp seals are seen along the Norwegian west coast during these migrations and sometimes even further south with individuals reaching the coasts of Denmark, Germany, the Netherlands and France (see Van Bree *et al.* 1994).

Besides the two above-mentioned age groups, six dates of Danish harp seals are scattered between ca. 3.3 cal. kyr BP and ca. AD 1000. They may represent single migration waves from the arctic as described

above, but they could also represent individuals stray-
ing from a small breeding population which, according to Storå & Lougas (2005), was probably
maintained in the Baltic during the Bronze Age and
the Iron Age. The causes for the final extinction of the
south Scandinavian breeding population is not
known.

Whales

Kinze (2007) has recently presented the modern Danish whale record based on sightings and strandings.
The survey can be compared with a list of whales
stranded or caught along the Danish coasts based on
historical records and going back to 1575 (Kinze
1995) and furthermore with the fossil record given
here.

If the whale fauna is defined as the species either
stranded on the Danish coasts or sighted in Danish
waters, the modern Danish whale fauna comprises 21
species. When adding up the number of records for
the period AD 1575–1991, the following six species get
the highest score: harbour porpoise *P. phocoena*,
white-beaked dolphin *L. albirostris*, killer whale *O.
orca*, common minke whale *B. acutorostrata*, bottlenose dolphin *T. truncatus* and sperm whale *P. macrocephalus* (Kinze 1995). Only 15 different species are
known from the subfossil record, but it is interesting
to see that the same six species also have the highest
frequency of occurrence in the past. However, one
species which is totally without a modern or historical
record, the bowhead or Greenland right whale, *B.
mysticetus* also belong to this group with dated records
ranging between ca. 17 cal. kyr BP and ca. AD 500. The
remaining 15 species belonging to the modern fauna
have only been registered a few times (<15) and eight
of these species recur in the subfossil list.

If the modern fauna is defined as the species
actually breeding in Danish waters and/or species
having the Danish part of the North Sea as part of
their natural geographical distribution, then the
modern whale fauna consists of six species: harbour
porpoise, white-beaked dolphin, Atlantic white-sided
dolphin *Lagonorhynchus acutus*, killer whale, long-
finned pilot whale *Globicephala melas* and common
minke whale.

Surprisingly, two of these whales, the white-sided
dolphin and the pilot whale, have no subfossil record
in Denmark, although a complete skeleton of the former has been found in a shell mound dated to the
Atlantic period at Otterö in Bohuslän on the Swedish
west coast (Lepiksaar 1966). The lack of subfossil evidence is probably due to the fact that they are both
offshore species normally found in deeper waters in

the Danish part of the North Sea and Skagerak; a low
number of strandings along the coast of the inner
Danish waters in the past should therefore be expected
(Aaris-Sørensen *et al.* in press).

Kinze (1991, 1995) characterizes the modern Danish whale fauna as a 'diluted' version of the North
Atlantic fauna where the British Isles seem to filter out
a number of species and where the main gateway to
the Danish waters runs through the Norwegian Channel. The species occur with varying frequency in Danish waters depending on their abundance in the
adjacent North Sea waters, and the occurrence of
oceanic and warm and cold water species is linked to
changes in climate and hydrographic phenomena.
Looking at the species composition and the distribution in time and space of the subfossil whales this
characterization equally applies to the prehistoric
whale fauna as well.

Conclusions

Due to fundamental differences in the taphonomic
pathways and in the accuracy of the different dating
methods used in the pre- and post-LGM fossil assemblages, differences are also found in space–time resolu-
tion of the available data sets. Thus, the history of the
Danish mammals up to and including the LGM has
to be painted with a much broader brush than the better founded Late- and Post Glacial part. With this in
mind the present investigation allows the following
conclusions:

- Only nine terrestrial mammals have been recorded
  from the Weichselian prior to LGM. Except for the
  two lemmings, *L. lemmus* and *D. torquatus*, they
  are all large herbivores, namely *M. primigenius*, *E.
  ferus*, cf. *C. antiquitatis*, *M. giganteus*, *R. tarandus*,
  B. priscus and *O. moschatus*. However, if the first
  appearing mammals *after* LGM are regarded as a
  last northward expansion of the European mammoth steppe fauna, they imply that at least *L. timidus*, *M. oeconomus*, *M. gregalis*, *C. lupus*, *V.
  lagopus*, *U. maritimus*, *M. erminea*, *M. nivalis*, *G.
  gulo* and *S. tatarica* should be included in the pre-
  LGM fauna.
- The fossil record shows that the range extensions and contractions of the mammalian species in southern Scandinavia are in accordance
  with the glacial history and palaeogeography as
  outlined by recent geological studies. The
  dynamics of the fauna match the advances and
  retreats of the ice cap and are best explained as
  a combination of expansions and local extinctions of marginal populations. Expansions of the

European mammoth steppe fauna into southern Scandinavia took place during all Weichselian interstadials.

- The diversity *after* LGM was characterized by a dramatic increase in richness between 15 and 11 cal. kyr BP followed by a decline towards a more moderate level between 11 and 9 kyr and then by a steady state the last 9000 years with a mean richness of ca. 30 species. The diversity was regulated by two opposing processes, colonizations and extinctions, which also caused large changes in species composition; faunal turnover rates were 90–60% during the first 4000 years and 20–4% during the last 6000 years. Thus, the total number of species recorded in the region is 53 and only one (*C. lupus*) has been continuously present throughout the whole period.

- The increase in richness from 15 to 11 kyr BP follows a marked increase in temperature, precipitation and subsequent increase in biological productivity, with changes in vegetation from sparse pioneer communities to grass/heath/steppe and finally to an open forest. The timing of the mammalian colonizations indicates that the immigration of the individual species occurred almost instantaneously as soon as a favourable habitat had been formed. Thus, the expansion of herbivore populations was controlled by the rate of 'habitat migration' – the time lag in vegetational response to the climatic improvement. The fossil record from adjacent areas shows that most species were present close to the southern Scandinavia border well in advance of the actual colonization.

- The Younger Dryas setback is registered by a drop in temperature, precipitation, biological productivity, vegetation and probably also by a local extinction around 12.6 cal. kyr BP of species such as *C. fiber*, *U. arctos*, *A. alces*, *M. giganteus*, *E. ferus* and possibly more. With the exception of *Megaloceros*, re-immigrations took place around 11.4 cal. kyr BP.

- The large number of species forming a peak during the 15–11 kyr interval reflects an overlap of early 'pioneers' in the form of species today characteristic of tundra and steppe environments with later colonizers in the form of species today characteristic of taiga and temperate forest. Such a mixture, containing species today considered as ecologically incompatible, is referred to as non-analogue or disharmonious assemblages. The composition of the non-analogue Late Glacial/Early Holocene mammal fauna is best explained as the result of individualistic responses by the species in accordance with their individual tolerance limits, in connection with environmental changes.

- The decline in richness between 11 and 9 kyr BP coincided with still rising temperatures and precipitation and a shift in vegetation from an open pioneer forest towards an ever denser mixed deciduous forest. The observed faunal turnover represents the exit of all arctic, subarctic and steppe species, and the arrival of species more adapted to forest environments

- A steady-state situation with species richness around 30 prevails from 9 kyr BP onwards. Temperature and precipitation were almost constant and a dense climax forest was established. No colonizations or extinctions occurred during the first 3000 years (the Atlantic period). During the last 6000 years the faunal turnover rate varied between 4% and 20% and it constantly changed the species composition. Man-made habitat fragmentation after the introduction of agriculture and husbandry and a steady increase in encounters with and persecutions from a growing human population led to extinctions. At the same time new species were attracted by the new open cultural landscape and/or the human habitations and activities.

- Denmark was transformed into a group of islands and peninsulas around 8 kyr BP. Shortly after, some of the larger islands reveal an impoverished fauna compared with the mainland (Jylland). This is believed to be caused by a higher vulnerability to climatic or biotic changes as water barriers prevent or reduce new immigration after population fluctuations. The changes are best documented on the largest island, Sjælland, with a local extinction during the 8–7 kyr interval of *U. arctos*, *M. putorius*, *M. meles*, *L. lynx*, *A. alces* and *B. primigenius*. The richness is regained during historical times due to the facts that: (1) *Mustela* and *Meles* reappear on the island during the 2–0 kyr interval; (2) the mainland suffers the same extinctions of *Bos*, *Alces*, *Ursus* and *Lynx*; and (3) Sjælland is affected by the same new colonizations as the mainland.

- The marine mammals are represented by 15 whale species, five seals and the polar bear. The record suggests that besides *U. maritimus*, *P. hispida* and *D. leucas* also *O. rosmarus*, *P. groenlandica*, *L. albirostris*, *O. orca*, *P. macrocephalus*, *B. acutorostrata*, *F. glacialis* and *B. mysticetus* have been present, at least occasionally, during all marine stages of the Weichselian. The oldest dates are confined to northern Jylland in agreement with the geographical distribution of the arctic–subarctic palaeo-Kattegat–Skagerrak. The youngest, on the other hand, are more widespread and much further south, in accordance with the Mid–Late Holocene transgressions creating the Inner Danish Waters. The younger occurrences include warm-adapted species such as *D. delphis*, *T. truncatus* and *Stenella* sp.

Acknowledgements. – First of all I wish to thank my colleagues Inge B. Enghoff, Anne Birgitte Gotfredsen, Tove Hatting, Jeppe Møhl, Knud Rosenlund and Ingrid Sørensen at the Quaternary Zoology Collections at the Zoological Museum (National History Museum of Denmark, University of Copenhagen) for years of joint collecting and/or managing the very large number of sub-fossil bone remains which form the basis of this investigation. I am most grateful for help given me by colleagues within other disciplines of Quaternary science especially Michael Houmark-Nielsen (geology and palaeogeography), Morten Fischer Mortensen and Bent Odgaard (palaeobotany) and Erik Brinch Petersen (archaeology). Special thanks are due to Alan Haywood and Paul Valdes for kindly providing me with palaeoclimate data sets from the BRIDGE Earth System Modelling Results. Knud Rosenlund is thanked for preparing drawings, figures and tables, Geert Brovad for taking all the photographs and Kristian M. Gregersen and Carsten Rabek for valuable comments on parts of the manuscript. The two reviewers, Wighart von Koenigswald (Bonn) and Derek Yalden (Manchester), are thanked for their valuable comments and suggestions on the manuscript. Finally, the Carlsberg Foundation is thanked for financial support in connection with the publication and the Aage V. Jensen Foundations for a most enjoyable and productive stay in their guest house in Imperia, Italy during spring 2007 which offered the necessary period of continuous peace to set the work going.

References

Aaris-Sørensen, K. 1976: A zoological investigation of the bone material from Sværdborg I-1943. *In* Henriksen, B.B. & Sværdborg, I. (eds): *Arkæologiske Studier III*, 137–148, Akademisk Forlag, København.

Aaris-Sørensen, K. 1980: Depauperation of the mammalian fauna of the island of Zealand during the Atlantic period. *Videnskabelige Meddelelser fra Dansk Naturhistorisk Forening 142*, 131–138.

Aaris-Sørensen, K. 1985: Den terrestriske pattedyrfauna i det sydfynske øhav gennem Atlantikum og Tidlig Subboreal. *In* Skaarup, J. (ed.): *Yngre Stenalder på øerne syd for Fyn*, 458–466. Langelands Museum, Rudkøbing.

Aaris-Sørensen, K. 1992: Deglaciation chronology and re-immigration of large mammals. A South Scandinavian example from Late Weichselian–Early Flandrian. *Courier Forschungsinstitut Senckenberg 153*, 143–149.

Aaris-Sørensen, K. 1995: Palaeoecology of a Late Weichselian vertebrate fauna from Nørre Lyngby, Denmark. *Boreas 24*, 355–365.

Aaris-Sørensen, K. 1998: *Danmarks Forhistoriske Dyreverden*, 3rd edn, 232 pp. Gyldendal, København.

Aaris-Sørensen, K. 1999: The Holocene history of the Scandinavian aurochs (*Bos primigenius* Bojanus, 1827). *Wissenschaftliche Schriften des Neanderthal Museums 1*, 49–57.

Aaris-Sørensen, K. 2003: Første vildhest fra dansk istid. Vildhesteknoglen fra Arnitlund. *Aktuel Arkæologi 3*, 6–7.

Aaris-Sørensen, K. 2004: Danmarks sidste urokser. *Aktuel Arkæologi 4*, 5–7.

Aaris-Sørensen, K. 2006: Northward expansion of the Central European megafauna during late Middle Weichselian interstadials, c. 45–20 kyr BP. *Palaeontographica Abt A 278*, 125–133.

Aaris-Sørensen, K. 2007: Fra istid til nutid (Late and Post Glacial mammals in Denmark). *In* Baagøe, H.J. & Jensen, T.S. (eds): *Dansk Pattedyratlas*, 312–321. Gyldendal, København.

Aaris-Sørensen, K. & Andreasen, T.N. 1995: Small mammals from Danish Mesolithic sites. *Journal of Danish Archaeology 11*(1992–93), 30–38.

Aaris-Sørensen, K. & Liljegren, R. 2004: Late Pleistocene remains of giant deer (*Megaloceros giganteus* Blumenbach) in Scandinavia: chronology and environment. *Boreas 33*, 61–73.

Aaris-Sørensen, K. & Petersen, K.S. 1984: A Late Weichselian find of polar bear (*Ursus maritimus* Phipps) from Denmark and reflections on the paleoenvironment. *Boreas 13*, 29–33.

Aaris-Sørensen, K., Petersen, K.S. & Tauber, H. 1990: Danish Finds of mammoth (*Mammuthus primigenius* (Blumenbach)). Stratigraphical position, dating and evidence of Late Pleistocene environment. *Danmarks Geologiske Undersøgelse Series B, 14*, 44.

Aaris-Sørensen, K., Petersen, K.S. & Henriksen, M.B. 1999: Late Weichselian record of Saiga (*Saiga tatarica* (L.)) from Denmark and its indications of glacial history and environment. *Quartär 33*, 87–94.

Aaris-Sørensen, K., Mühldorff, R. & Petersen, E.B. 2007: The Scandinavian reindeer (*Rangifer tarandus* L.) after the last glacial maximum: time, seasonality and human exploitation. *Journal of Archaeological Science 34*, 914–923.

Aaris-Sørensen, K., Rasmussen, K.L., Kinze, C. & Petersen, K.S. in press: Late Pleistocene and Holocene whale remains (Cetacea) from Denmark and adjacent countries: species, distribution, chronology and trace element concentrations. *Marine Mammal Science 26*, in press.

Andersen, S.H. 1993: Bjørnsholm. A stratified Køkkenmødding on the central Limfjord, North Jutland *Journal of Danish Archaeology 10*(1991), 59–96.

Andersen, S.Th. 1995: History of vegetation and agriculture at Hassing Huse Mose, Thy, Northwest Denmark, since the Ice Age. *Journal of Danish Archaeology 11*(1992–93), 57–79.

Andersen, S.H. 2008: The Mesolithic–Neolithic transition in western Denmark seen from a kitchen midden perspective: a survey. *Analecta Praehistorica Leidensia 40*, 67–74.

Andreasen, T.N. 1997: Taxonomic status of Desmana (Insectivora) and Spermophilus (Rodentia) specimens from Danish Late Weichselian deposits. *Acta Zoologica Cracoviensia 40*, 229–236.

Araújo, M.B., Nogués-Bravo, D., Diniz-Filho, J.A.F., Haywood, A.M., Valdes, P.J. & Rahbek, C. 2008: Quaternary climate changes explain diversity among reptiles and amphibians. *Ecography 31*, 8–15.

Arppe, L. & Karhu, J.A. in press: Oxygen isotope values of precipatation and thermal climate in Europe during the Middle to Late Weichselian ice age. *Quaternary Science Reviews* (in press).

Baagøe, H.J. 2001: Danish bats (Mammalia: Chiroptera): atlas and analysis of distribution, occurrence, and abundance. *Steenstrupia 26*, 1–117.

Baagøe, H.J. 2007: Vandflagermus, *Myotis daubentonii* (Kuhl, 1817). *In* Baagøe, H.J. & Jensen, T.S. (eds): *Dansk Pattedyratlas*, 56–59. Gyldendal, København.

Baales, M. 2001: From lithics to spatial and social organization: interpreting the lithic distribution and raw material composition at the final palaeolithic site of Kettig (central Rhineland, Germany). *Journal of Archaeological Science 28*, 127–141.

Baales, M. & Street, M. 1996: Hunter-gatherer behaviour in a changing Late Glacial landscape: Allerød Archaeology in the central Rhineland, Germany. *Journal of Anthropological Research 52*, 281–316.

Behling, H. & Street, M. 1999: Palaeoecological studies at the Mesolithic site at Bedburg–Königshoven near Cologne, Germany. *Vegetation History and Archaeobotany 8*, 273–285.

Benecke, N. 2000: Die Jungpleistozäne und Holozäne Tierwelt Mecklenburg-Vorpommerns. *Beiträge zur Ur- und Frühgeschichte Mitteleuropas 23*, 1–143.

Benecke, N. 2004: Faunal succession in the lowlands of Northern Central Europe at the Pleistocene–Holocene transition. *In* Terberger, T. & Eriksen, B.V. (eds): Hunters in a changing world. Environment and archaeology of the Pleistocene–Holocene transition (ca. 11000–9000 BC) in northern central Europe. *Internationale Archäologie 5*, 43–52.

Benecke, N., Gramsch, B. & Weisse, R. 2002: Zur Neudatierung des Ur-Fundes von Potsdam-Schlaatz, Brandenburg. *Archäologisches Korrespondenzblatt 32*, 161–168.

Bennike, O., Houmark-Nielsen, M., Böcher, J. & Heiberg, E.O. 1994: A multi-disciplinary macrofossil study of Middle Weichselian sediments at Kobbelgård, Møn, Denmark. *Palaeogeography, Palaeoclimatology, Palaeoecology 111*, 1–15.

Bennike, O., Rasmussen, P. & Aaris-Sørensen, K. 2008: The harp seal (*Phoca groenlandica* Erxleben) in Denmark, southern Scandinavia, during the Holocene. *Boreas 37*, 263–272.

Berglund, B.E., Håkansson, S. & Lagerlund, E. 1976: Radiocarbon-dated mammoth (*Mammuthus primigenius* Blumenbach) finds in South Sweden. *Boreas 5*, 177–191.

Berglund, B.E., Håkansson, S. & Lepiksaar, J. 1992: Late Weichselian polar bear (*Ursus maritimus* Phipps) in southern Sweden. *Sveriges Geologiska Undersökning, Ser. Ca 81*, 31–42.

Björck, S. 1995: A review of the history of the Baltic Sea, 13.0–8.0 ka BP. *Quaternary International 27*, 19–40.

Björck, S., Ekström, J., Iregren, E., Larsson, L. & Liljegren, R. 1996: Reindeer and palaeoecological and palaeogeographic changes in South Scandinavia during late-glacial and early post-glacial times. *Arkæologiske Rapporter, Esbjerg Museum 1*, 195–214.

Björck, S., Walker, M.J.C., Cwynar, L.C., Johnsen, S., Knudsen, K.-L., Lowe, J.J., Wohlfarth, B. & INTIMATE Members 1998: An event stratigraphy for the last Termination in the North Atlantic region based on the Greenland ice-core record: a proposal by the INTIMATE group. *Journal of Quaternary Science 13*, 283–292.

Blystad, P., Thomsen, H., Simonsen, A. & Lie, R.W. 1983: Find of a nearly complete Late Weichselian polar bear skeleton, *Ursus maritimus* Phipps, at Finnøy, southwestern Norway: a preliminary report. *Norsk Geologisk Tidsskrift 63*, 193–197.

Bondesen, P. & Lykke-Andersen, H. 1978: The desman, *Desmana moschata* (L.) – a new mammal in Denmark after the Ice Age. *Natura Jutlandica 20*, 25–32.

Borgen, U. 1979: Ett fynd av fossil myskoxe i Jämtland och något om myskoxarnas biologi och historia. *Fauna och flora, Årgang 74*, 1–12.

Bratlund, B. 1993: The bone remains of mammals and birds from the Bjørnsholm Shell-Mound. A preliminary report. *Journal of Danish Archaeology 10*(1991), 97–104.

Bratlund, B. 1999: A revision of rarer species from the Ahrensburgian assemblage of Stellmoor. *In* Benecke, N. (ed.): *The Holocene History of the European Vertebrate Fauna. Archäologie in Eurasien Band 6*, 39–42. Verlag Marie Leidorf, Rahden/Westfahlen.

Bronk Ramsey, C. 1995: Radiocarbon calibration and analysis of stratigraphy: the OxCal program. *Radiocarbon 37*, 425–430.

Bronk Ramsey, C. 2001: Development of the radiocarbon calibration program OxCal. *Radiocarbon 43*, 255–363.

Bronk Ramsey, C., Higham, T.F.G., Owen, D.C., Pike, A.W.G. & Hedges, R.E.M. 2002: Radiocarbon dates from the Oxford AMS system: archaeometry datelist 31. *Archaeometry 44*(Suppl. 1), 1–149.

Brothwell, D. 1981: The Pleistocene and Holocene archaeology of the house mouse and related species. *Symposium of the Zoological Society of London 47*, 1–13.

Brown, J.H., Ernest, S.K.M., Parody, J.M. & Haskell, J.P. 2001: Regulation of diversity: maintenance of species richness in changing environments. *Oecologia 126*, 321–332.

Christensen, C. 1982: Havniveauændringer 5500–2500 f. Kr. i Vedbækområdet, NØ-Sjælland. *Dansk Geologisk Forenings Årsskrift 1981*, 91–107.

Christiansen, T.S. & Jensen, T.S. 2007: Pattedyr på øerne. *In* Baagøe, H.J. & Jensen, T.S. (eds): *Dansk Pattedyratlas*, 338–342. Gyldendal, København.

Clausen, I. 1996: Alt Düvenstedt LA 121, Schleswig–Holstein. – occurrence of the Ahrensburgian culture in soils of the Alleröd interstadial. A preliminary report. *In* Larsson, L. (ed.): The earliest settlement of Scandinavia and its relationship with neighbouring areas. *Acta Archaeologica Lundensia, Series in 8°, 24*, 99–110.

Coope, G.R. & Lister, A.M. 1987: Late-glacial mammoth skeletons from Condover, Shropshire, England. *Nature 330*, 472–474.

Cordy, J.-M. 1991: Palaeoecology of the Late Glacial and early Postglacial of Belgium and neighbouring areas. *In* Barton,

R.N.E., Roberts, A.J. & Roe, D.A. (eds): The late glacial in north-west Europe. Human adaptation and environmental change at the end of the Pleistocene. *CBA Research Report 77*, 40–47.

Crégut-Bonnoure, E. 1992: Dynamics of Bovid migration in western Europe during the Middle and Late Pleistocene. *Courier Forschungsinstitut Senckenberg 153*, 177–185.

Crégut-Bonnoure, E. & Gagnière, S. 1981: Sur la presence de *Saiga tatarica* (Mammalia, Artiodactyla) dans le depot Pléistocène supérieur de la grotte de la Salpétrière à Remoulins (Gard, France). *Nouvelles archives du Muséum d'histoire naturelle de Lyon 19*, 37–42.

Currant, A.P. 1987: Late Pleistocene saiga antelope *Saiga tatarica* on Mendip. *Proceedings University Bristol Spelaeological Society 18*, 74–80.

Currant, A.P., Jacobi, R.M. & Stringer, C.B. 1989: Excavations at Gough's Cave, Somerset 1986–7. *Antiquity 63*, 131–136.

Danell, K., Lundberg, P. & Niemelä, P. 1996: Species richness in mammalian herbivores: patterns in the boreal zone. *Ecography 19*, 404–409.

Davidsen, K. 1978: The final TRB culture in Denmark. A settlement study. *Arkæologiske Studier, V*, 1–207.

Degerbøl, M. 1928: Mindre Bidrag til Danmarks forhistoriske Dyreverden. I. Et Fragment af Bæver (*Castor fiber* L.) fra Bronzealderen. *Videnskabelige Meddelelser fra Dansk Naturhistorisk Forening 86*, 75–81.

Degerbøl, M. 1932: Et Fund af Steppe-Antilope (*Saiga tatarica* (Pall.)) i Danmark. *Meddelelser fra Dansk Geologisk Forening 8*, 175–184.

Degerbøl, M. 1933: Danmarks Pattedyr i Fortiden i Sammenligning med recente Former. I. *Videnskabelige Meddelelser fra Dansk Naturhistorisk Forening 96*, 357–641.

Degerbøl, M. 1942: Et knoglemateriale fra Dyrholm-Bopladsen, en Ældre Stenalder-Køkkenmødding med særligt henblik på uroksens køns-dimorphisme og paa kannibalisme i Danmark. *Det Kongelige Danske Videnskabernes Selskab. Arkæologisk-Kunsthistoriske Skrifter I 1*, 77–135.

Degerbøl, M. 1943: Om dyrelivet i Aamosen i stenalderen. *In* Mathiassen, Th. (ed.): Stenalderbopladser i Aamosen. *Nordiske Fortidsminder III, 3*, 165–206.

Degerbøl, M. 1946a: Dyreknogler fra en senglacial Boplads ved Bromme. *In* Mathiassen, Th. (ed.): En senglacial Boplads ved Bromme. *Aarbøger for Nordisk Oldkyndighed og Historie 1946*, 137–147.

Degerbøl, M. 1946b: Dyreknogler fra Borggraven ved Søborg Slot. Træk af Husdyrholdet i Middelalderen. *Fra det gamle Gilleleje 1946*, 77–88.

Degerbøl, M. 1964: Some remarks on Late- and Post-glacial vertebrate fauna and its ecological relations in northern Europe. *Journal of Animal Ecology 33*(Suppl.), 71–85.

Degerbøl, M. & Fredskild, B. 1970: The Urus (*Bos primigenius* Bojanus) and Neolithic domesticated cattle (*Bos taurus domesticus* Linné) in Denmark. *Det Kongelige Danske Videnskabernes Selskab, Biologiske Skrifter 17*(1), 1–234.

Degerbøl, M. & Krog, H. 1959: The reindeer (*Rangifer tarandus* L.) in Denmark. *Biologiske Skrifter udgivet af Det Kongelige Danske Videnskabernes Selskab 10*(4), 1–165.

Delpech, F. 1983: Les faunes du Paléolithique supérieur dans le sud-ouest de la France. *Cahiers du Quaternaire 6*, 1–453.

Ekström, J. 1993: The Late Quaternary history of the Urus (*Bos primigenius* Bojanus 1827) in Sweden. *LUNDQUA Thesis 29*, 1–129.

Ellis, C.J., Allen, M.J., Gardiner, J., Harding, P., Ingrem, C., Powell, A. & Scaife, R.G. 2003: An Early Mesolithic seasonal hunting site in the Kennet Valley, southern England. *Proceedings of the Prehistoric Society 69*, 107–135.

Enghoff, I.B. 1984: Nyt gammelt fund af desmanen i Danmark. *Dyr i Natur og Museum 2*, 14–16.

Eriksson, M. & Magnell, O. 2001: Det djuriska Tågerup. *In* Karsten, P. & Knarrström, B. (eds): *Tågerup Specialstudier*, 157–237. Riksantikvarieämbetet, Lund.

Erslev E. 1871: *Om de Glubende Dyrs Undergang i Nørrejylland*, 32 pp. Kjøbenhavn.

Fischer, A. 1990: A late Palaeolithic flint workshop at Egtved, east Jutland – a glimpse of the Federmesser culture in Denmark. *Journal of Danish Archaeology* 7(1988), 7–23.

Fischer, A. & Tauber, H. 1987: New C-14 datings of Late Palaeolithic cultures from northwestern Europe. *Journal of Danish Archaeology* 5(1986), 7–13.

Fisher, C.T. & Yalden, D.W. 2004: The steppe pika *Ochotona pusilla* in Britain, and a new northerly record. *Mammal Review 34*, 320–324.

Fjellberg, A. 1978: Fragments of a Middle Weichselian fauna on Andøya, north Norway. *Boreas 7*, 39.

Fraser, F.C. & King, J.E. 1954: Faunal remains. *In* Clark, J.G.D. (ed.): *Excavations at Star Carr*, 70–95. Cambridge University Press, Cambridge.

Fredén, C. 1975: Subfossil finds of arctic whales and seals in Sweden. Appendix: radiocarbon determinations of miscellaneous subfossil finds of the Swedish west coast. *Sveriges Geologiska Undersökning C 710*, 62.

Fredén, C. 1984: Faunahistoriska notiser om några av Naturhistoriska Muséets daterade subfossila fynd. *Göteborgs Naturhistoriska Museum Årstryck 1984*, 31–45.

Gehl, O. 1961: Die Säugetiere. *In* Schuldt, E. (ed.): Hohen Viecheln. Ein mittelsteinzeitlicher Wohnplatz in Meclenburg. *Schriften der Sektion für Vor- und Frühgeschichte 10*, 40–63.

Gelskov, S.V. 2005: Dyreknogler fra voldstederne på Langeland og Ærø. *In* Skaarup, J. (ed.): *Øhavets Middelalderlige Borge og Voldsteder*, 454–477. Langelands Museum, Rudkøbing.

Gjøsæter, H., Bogstad, B. & Tjelmeland, S. 2003: *News 2003*, 3 pp. Institute of Marine Research, Bergen.

Graham, R.W. & Semken, H.A. 1987: Philosophy and procedures for Paleoenvironmental studies of Quaternary mammalian faunas. *In* Graham, R.W., Semken, H.A. & Graham, M.A. (eds): Late Quaternary mammalian biogeography of the Great Plains and Prairies. *Illinois State Museum Scientific Papers 22*, 1–17.

Graham, R.W., Lundelius Jr., E.L., Graham, M.A., Schroeder, E.K., Toomey III, R.S., Anderson, E., Barnosky, A.D., Burns, J.A., Churcher, C.S., Grayson, D.K., Guthrie, R.D., Harington, C.R., Jefferson, G.T., Martin, L.D., McDonald, H.G., Morlan, R.E., Semken Jr., H.A., Webb, S.D., Werdelin, L. & Wilson, M.C. of the FAUNMAP Working Group 1996: Spatial response of mammals to Late Quaternary environmental fluctuations. *Science 272*, 1601–1606.

Gramsch, B. 2000: Friesack: Letzte Jäger und Sammler in Brandenburg. *Jahrbuch des Römisch-Germanischen Zentralmuseums Mainz 47*, 51–96.

Grimm, S.B. & Weber, M.-J. 2008: The chronological framework of the Hamburgian in the light of old and new ^{14}C-dates. *Quartär 55*, 17–40.

Hansen, K.M., Petersen, E.B. & Aaris-Sørensen, K. 2004: Filling the gap: Early Preboreal Maglemose elk deposits at Lundby, Sjælland, Denmark. *In* Terberger, T. & Eriksen, B.V. (eds): Hunters in a changing world. Environment and archaeology of the Pleistocene–Holocene transition (c. 11,000–9,000 BC) in northern central Europe. *Internationale Archäologie 5*, 75–84.

Hatting, T. 1995: Det er ikke alt guld, der glimter – et uroksehorn fra Eskær Mose. *Vendsyssel Nu & Da 15*, 22–25.

Hawkins, B.A. & Porter, E.E. 2003: Relative influences of current and historical factors on mammal and bird diversity patterns in deglaciated North America. *Global Ecology and Biogeography 12*, 475–481.

Hawkins, B.A., Field, R., Cornell, H.V., Currie, D.J., Guégan, J.-F., Kaufman, D.M., Kerr, J.T., Mittelbach, G.G., Oberdorff, T., O'Brian, E.M., Porter, E.E. & Turner, J.R.G. 2003: Energy, water, and broad-scale geographic patterns of species richness. *Ecology 84*, 3105–3117.

Hedges, R.E.M., Housley, R.A., Bronk Ramsey, C. & van Klinken, G.J. 1994: Radiocarbon dates from the Oxford AMS system: Archaeometry datelist 18. *Archaeometry 36*, 337–374.

Hedges, R.E.M., Housley, R.A., Pettitt, P.B., Bronk Ramsey, C. & van Klinken, G.J. 1996: Radiocarbon dates from the Oxford AMS system: Archaeometry datalist 21. *Archaeometry 38*, 181–207.

Heiberg, E.O. 1995: Sen- og Postglaciale mindre gnavere (Rodentia) og insektædere (Insectivora) fra Danmark, 273 pp. Unpublished MSc thesis, Geological Institute, University of Copenhagen.

Heiberg, E.O. & Bennike, O. 1997: Late Quaternary Rodents from Southwestern Baltic Sea. *Baltica 10*, 47–52.

Heinrich, D. 1976: Bemerkungen zum mittelalterlichen Vorkommen der Wanderratte (*Rattus norvegicus* Berkenhout, 1769) in Schleswig–Holstein. *Zoologischer Anzeiger 196*, 273–278.

Hetherington, D.A., Lord, T.C. & Jacobi, R.M. 2006: New evidence for the occurrence of Eurasian lynx (*Lynx lynx*) in medieval Britain. *Journal of Quaternary Science 21*, 3–8.

Hewitt, G. 2000: The genetic legacy of the Quaternary ice age. *Nature 405*, 907–913.

Hewitt, C.D., Senior, C.A. & Mitchell, J.F.B. 2001: The impact of dynamic sea-ice on the climatology and climate sensitivity of a GCM: a study of past, present, and future climates. *Climate Dynamics 17*, 655–668.

Holm, J. 1993: Settlements of the Hamburgian and Federmesser cultures at Slotseng, South Jutland. *Journal of Danish Archaeology 10*(1991), 7–19.

Holm, J. & Rieck, F. 1992: Istidsjægerne ved Jelssøerne. Hamburgkulturen i Danmark. *Skrifter fra Museumsrådet for Sønderjyllands Amt Haderslev 5*, 93–132.

Houmark-Nielsen, M. 2004: The Pleistocene of Denmark: a review of stratigraphy and glaciation history. *In* Ehlers, J. & Gibbard, P.L. (eds): *Quaternary Glaciations, Extent and Chronology, Part 1: Europe*, 35–46. Elsevier, Amsterdam.

Houmark-Nielsen, M. & Kjær, K.H. 2003: Southwest Scandinavia, 40–15 kyr BP: palaeogeography and environmental change. *Journal of Quaternary Science 18*, 769–786.

Houmark-Nielsen, M., Krüger, J. & Kjær, K.H. 2005: De seneste 150.000 år i Danmark. Istidslandskabet og naturens udvikling. *Geoviden – Geologi og Geografi 2*, 1–19.

Houmark-Nielsen, M., Knudsen, K.L. & Noe-Nygård, N. 2006: Istider og mellemistider. *In* Larsen, G. (ed.): *Naturen i Danmark. Geologien*, 255–302. Gyldendal, København.

Housley, R.A. 1991: AMS dates from the Late Glacial and early Postglacial in north-west Europe: a review. *In* Barton, N., Roberts, A.J. & Roe, D.A. (eds): The Late Glacial in north-west Europe: human adaption and environmental change at the end of the Pleistocene. *CBA Research Report 77*, 25–39.

Hufthammer, A.K. 2001: The Weichselian (c. 115,000–10,000 BP) vertebrate fauna of Norway. *Bollettino della Società Paleontologica Italiana 40*, 201–208.

Iversen, J. 1942: En pollenanalytisk Tidsfæstelse af ferskvandslagene ved Nørre Lyngby. *Meddelelser fra Dansk Geologisk Forening 10*(2), 130–151.

Jansen, T., Forster, P., Levine, M.A., Oelke, H., Hurles, M., Renfrew, C., Weber, J. & Olek, K. 2002: Mitochondrial DNA and the origins of the domestic horse. *Proceedings of the National Academy of Sciences 99*(16), 10905–10910.

Jensen, P. 1993: Body size trends of Roe Deer (*Capreolus capreolus*) from Danish Mesolithic sites. *Journal of Danish Archaeology 10*(1991), 51–58.

Jensen, A. & Jensen, B. 1972: Ilderen (*Putorius putorius*) og ilderjagten i Danmark 1969/70. *Danske Vildtundersøgelser 21*, 1–32.

Jensen, T.S. & Møller, J.D. 2007: Birkemus, *Sicista betulina* (Pallas, 1779). *In* Baagøe, H.J. & Jensen, T.S. (eds): *Dansk Pattedyratlas*, 170–173. Gyldendal, København.

Jessen, K. 1935: The composition of the forest in Northern Europe in Epipalaeolithic time. *Det Kongelige Danske Videnskabernes Selskab, Biologiske Meddelser XII, 1*, 1–64.

Jessen, A. & Nordmann, V. 1915: Ferskvandslagene ved Nørre Lyngby. Summary: the freshwater deposits at Nørre Lyngby. *Danmarks Geologiske Undersøgelse II 29*, 1–66.

Joensen, A.H., Søndergaard, N.-O. & Hansen, E.B. 1976: Occurrence of seals and seal hunting in Denmark. *Danish Review of Game Biology 10*, 1–20.

Kahlke, R.-D. 1990: Der Saiga-Fund von Pahren. Ein Beitrag zur Kenntnis der paläarktischen Verbreitungsgeschichte der Gattung Saiga GRAY 1843 unter besonderer Berücksichtigung des Gebietes der DDR. *Eiszeitalter u. Gegenwart 40*, 20–37.

Kahlke, R.-D. 1992: Repeated immigration of saiga into Europe. *Courier Forschungsinstitut Senckenberg 153*, 187–195.

Kahlke, R.-D. 1999: *The History of the Origin, Evolution and Dispersal of the Late Pleistocene Mammuthus–Coelodonta Faunal Complex in Eurasia (Large Mammals)*, 219 pp. Fenske Companies, Rapid City, SD.

King, J.E. 1962: Report on animal bones. *In* Wymer, J. (ed.): Excavations at the Maglemosian Sites at Thatcham, Berkshire, England. *Proceedings of the Prehistoric Society 28*, 355–361.

Kinze, C. 1991: Hvaler. *In* Muus, B. (ed.): *Danmarks Pattedyr 2*, 104–128. Gyldendal, København.

Kinze, C. 1995: Danish whale records 1575–1991 (Mammalia, Cetacea). Review of whale specimens stranded, directly or incidentally caught along the Danish coasts. *Steenstrupia 21*, 155–196.

Kinze, C. 2007: Hvidnæse, *Lagenorhynchus albirostris* Gray, 1846; Almindelig delfin, *Delphinus delphis* Linnaeus, 1758; Øresvin, *Tursiops truncatus* (Montagu, 1821); Stribet delfin, *Stenella coeruleoalba* (Meyen, 1833); Spækhugger, *O. orca* (Linnaeus, 1758); Døgling, *Hyperoodon ampullatus* (Forster, 1770); Kaskelot, *Physeter macrocephalus* (Linnaeus, 1758); Vågehval, *Balaenoptera acutorostrata* (Lacépède, 1804); Finhval, *Balaenoptera physalus* (Linnaeus, 1758); Pukkelhval, *Megaptera novaeanglia* (Borowski, 1781); Nordkaper, *Eubalaena glacialis* (Müller, 1776). *In* Baagøe, H.J. & Jensen, T.S. (eds): *Dansk Pattedyratlas*, 264–311. Gyldendal, København.

Kjær, K.H., Lagerlund, E., Adrielsson, L., Thomas, P.J., Murray, A. & Sandgren, P. 2006: The first independent chronology for Middle and Late Weichselian sediments from southern Sweden and the Island of Bornholm. *GFF 128*, 209–219.

von Koenigswald, W. 2002: *Lebendige Eiszeit. Klima und Tierwelt im Wandel*, 190 pp. Wissenschaftliche Buchgesellschaft, Darmstadt.

von Koenigswald, W. 2003: Mode and causes for the Pleistocene turnovers in the mammalian fauna of central Europe. *Deinsea 10*, 305–312.

Kowalski, K. 2001: Pleistocene rodents of Europe. *Folia Quaternaria 72*, 3–389.

Krause, W. 1937: Die eisenzeitliche knochenfunde von Meiendorf. *In* Rust, A. (ed.): *Das Altsteinzeitliche Rentierjagerlager Meiendorf*, 48–61. Karl Wachholtz Verlag, Neumünster.

Krause, W. & Kollau, W. 1943: Die steinzeitlichen wirbeltierfaunen von stellmoor in Holstein. *In* Rust, A. (ed.): *Die alt- und Mittelsteinzeitlichen Funde von Stellmoor*, 49–59. Karl Wachholtz Verlag, Neumünster.

Kubiak, H. 1980: The skulls of *Mammuthus primigenius* (Blumenbach) from Debica and Bzianka near Rzeszów, South Poland. *Folia Quaternaria 51*, 31–45.

Lagerlund, E. & Houmark-Nielsen, M. 1993: Timing and pattern of the last deglaciation in the Kattegat region, southwest Scandinavia. *Boreas 22*, 337–347.

Larsson, L., Liljegren, R., Magnell, O. & Ekström, J. 2002: Archaeofaunal aspects of bog finds from Hässleberga, southern Scania, Sweden. *In* Eriksen, B.V. & Bratlund, B. (eds): *Recent Studies in the Final Palaeolithic of the European Plain*, 61–74. Jutland Archaeological Society, Højbjerg.

Lauritzen, S.-E., Nese, H., Lie, R.W., Lauritsen, Å. & Løvlie, R. 1996: Interstadial/interglacial fauna from Nordcemgrotta, Kjøpsvik, north Norway: climate change: the Karst record. *Extended Conference Abstracts: Karst Waters Institute Special Publication 2*, 89–92.

Lepiksaar, J. 1964: Subfossile Robbenfunde von der schwedischen Westküste. *Zeitschrift für Säugetierkunde 29*, 257–266.

Lepiksaar, J. 1966: Zahnwalfunde in Schweden. *Bijdragen tot de Dierkunde 36*, 3–16.

Lepiksaar, J. 1986: The holocene history of Theriofauna in Fennoscandia and Baltic countries. *Striae 24*, 51–70.

Lie, R.W. 1986: Animal bones from the Late Weichselian in Norway. *Fauna Norvegica, Serie A 7*, 41–46.

Lie, R.W. 1990: Blomvågfunnet, de eldste spor etter mennesker i Norge? *Viking 53*, 7–20.

Liljegren, R. & Ekström, J. 1996: The terrestrial late glacial fauna in south Sweden. *Acta Archaeologica Lundensia, Series in 8° 24*, 135–139.

Liljegren, R. & Lagerås, P. 1993: *Från Mammutstäpp till Kohage. Djurens Historia i Sverige*, 48 pp. Wallin & Dalholm, Lund.

Lister, A.M. 1989: Rapid dwarfing of red deer on Jersey in the Last Interglacial. *Nature 342*, 539–542.

Lister, A.M. 1991: Lateglacial mammoths in Britain. *In* Barton, N., Roberts, A.J. & Roe, D.A. (eds): The Late Glacial in north-west Europe: human adaption and environmental change at the end of the Pleistocene. *CBA Research Report 77*, 51–59.

Lister, A.M. 1994: The evolution of the giant deer, *Megaloceros giganteus* (Blumenbach). *Zoological Journal of the Linnean Society 112*, 65–100.

Lodal, J. 2007: Husrotte, *Rattus rattus* (Linnaeus, 1758). *In* Baagøe, H.J. & Jensen, T.S. (eds): *Dansk Pattedyratlas*, 156–159. Gyldendal, København.

Lougas, L. 1998: Postglacial invasions of the Harp Seal (*Pagophilus groenlandicus* Erxl. 1777) into the Baltic Sea. *Proceedings of the Latvian Academy of Sciences. Section B, 52*, 63–69.

MacArthur, R.H. & Wilson, E.O. 1967: *The Theory of Island Biogeography*, 203 pp. Princeton University Press, Princeton, NJ.

Mangerud, J. 1977: Late Weichselian marine sediments containing shells, foraminifera and pollen, at Ågotnes, western Norway. *Norsk Geologisk Tidsskrift 57*, 23–57.

Matthews, J.A. 1992: *The Ecology of Recently-Deglaciated Terrain – A Geological Approach to Glacier Forelands and Primary Succession*, 386 pp. Cambridge University Press, Cambridge.

Mitchell-Jones, A.J., Amori, G., Bogdanowicz, W., Krystufek, B., Reijnders, P.J.H., Spitzenberger, F., Stubbe, M., Thissen, J.B.M., Vohralík, V. & Zima, J. 1999: *The Atlas of European Mammals*, 484 pp. Academic Press, London.

Møbjerg, T. & Rostholm, H. 2006: Foreløbige resultater af de arkæologiske undersøgelser ved Bølling Sø. *In* Eriksen, B.V. (ed.): Stenalderstudier. Tidligt Mesolitiske Jægere og Samlere i Sydskandinavien. *Jysk Arkæologisk Selskabs Skrifter 55*, 147–159.

Møhl, U. 1971a: Fangstdyrene ved de danske strande. Den zoologiske baggrund for harpunerne. *KUML. Årbog for Jysk arkæologisk Selskab 1970*, 297–329.

Møhl, U. 1971b: Et knoglemateriale fra Vikingetid og Middelalder i Århus. *In* Andersen, H.H., Crabb, P.J. & Madsen, H.J. (eds): Århus Søndervold. *Jysk Arkæologisk Selskabs Skrifter IX*, 321–329.

Møhl, U. 1977: Bjørnekløer og brandgrave. Dyreknogler fra germansk jernalder i Stilling. KUML. *Årbog for Jysk arkæologisk Selskab 1977*, 119–129.

Møhl, U. 1985: The walrus, *Odobenus rosmarus* (L.), as a "Danish" faunal element during the Weichselian Ice Age. *Bulletin of the Geological Society of Denmark 34*, 83–85.

Mortensen, M.F. 2007: Biostratigraphy and chronology of the terrestrial late Weichselian in Denmark – new investigations of the vegetation development based on pollen and plant macrofossils in the Slotseng basin. Unpublished PhD thesis. University of Aarhus.

Mortensen, M.F., Holm, J., Christensen, C. & Aaris-Sørensen, K. 2008: De første mennesker i Danmark. *Nationalmuseets Arbejdsmark 2008*, 69–82.

Motuzko, A.N. & Ivanov, D.L. 1996: Holocene micromammal complexes of Belerus: a model of faunal development during Interglacial epochs. *Acta Zoologica Cracoviensia 39*, 381–386.

Munthe, H. 1905: Om ett Fynd af Kvartär Myskoxe vid Nol i Bohuslän. *Sveriges Geologiska Undersökning, Serie C, No 197*, 3–19/Geologiska Föreningens Förhandlingar 27, 173–189.

Nordmann, V. 1944: *Jordfundne Pattedyrlevninger I Danmark*, 112 pp. H. Hagerups Forlag, København.

Nordmann, V. & Degerbøl, M. 1930: En fossil Kæbe af Isbjørn (*Ursus maritimus* L.) fra Danmark. *Videnskabelige Meddelelser fra Dansk Naturhistorisk Forening 88*, 273–286.

Odgaard, B. 2006: Fra bondestenalder til nutid. *In* Larsen, G. (ed.): *Naturen i Danmark. Geologien*, 333–359. Gyldendal, København.

Petersen, B.F. 2001: Senpalæolitiske opsamlingsfund fra Sydsjælland, Fejø og Nordsjælland – et bidrag til udforskningen af de senglaciale kulturer i Danmark. *Kulturhistoriske Studier, Sydsjællands Museum 2001*, 7–64.

Petersen, K.S. 2004: Late Quaternary environmental changes recorded in the Danish marine molluscan faunas. *Geological Survey of Denmark and Greenland Bulletin 3*, 1–268.

Petersen, P.V. 2006: White Flint and Hilltops – Late Palaeolithic finds in southern Denmark. *In* Hansen, K.M. & Pedersen, K.B. (eds): Across the Western Baltic. *Sydsjællands Museums Publikationer 1*, 57–74.

Petersen, E.B. in press: The human settlement of southern Scandinavia 12 500–8 700 cal BC. *In* Street, M., Barton, R.N.E. & Terberger, T. (eds): *Humans, Environment and Chronology of the Late Glacial of the North European Plain*. Proceedings of Workshop 14 (Commission XXXII "The Final Palaeolithic of the Great European Plain) of the 15th U. I. S. P. P. Congress, Lisbon, September 2006. RGZM-Tagungen Band, Mainz.

Petersen, E.B. & Egeberg, T. in press: From dragsholm I to dragsholm II. *In* Lübke, H., Lüth, F. & Terberger, T. (eds): *Neue Forschungen zum Gräberfeld von Ostorf*. RGZM-Tagungen Band, Mainz.

Petersen, P.V. & Johansen, L. 1993: Sølbjerg I – an Ahrensburgian site on a reindeer migration route through eastern Denmark. *Journal of Danish Archaeology 10*(1991), 20–37.

Raahauge, T.N. 2002: Fauna and cultural landscape in Thy during the transition of Subboreal and Subatlantic. A palaeozoological regional analysis of fauna, husbandry and landscape based on bone remains from Bronze Age and Early Iron Age settlements. Unpublished PhD thesis, Zoological Museum, University of Copenhagen, 106 pp.

Rasmussen, S.O., Andersen, K.K., Svensson, A.M., Steffensen, J.P., Vinther, B.M., Clausen, H.B., Siggaard-Andersen, M.-L., Johnsen, S.J., Larsen, L.B., Dahl-Jensen, D., Bigler, M., Röthlisberger, R., Fischer, H., Goto-Azuma, K., Hansson, M.E. & Ruth, U. 2006: A new Greenland ice core chronology for the last glacial termination. *Journal of Geophysical Research 111*, D06102, doi: 10.1029/2005JD006079.

Rasmussen, P., Nielsen, A.B. & Bradshaw, E. 2007: Fra natur- til kulturlandskab. *Geoviden – Geologi og Geografi 1*, 19.

Raufuss, I. & von Koenigswald, W. 1999: New remains of the Pleistocene *Ovibos moschatus* from Germany and its geographic and stratigrapic occurrence in Europe. *Geologie en Mijnbouw 78*, 383–394.

Requate, H. 1956: Die Jagdtiere in den Nahrungsresten einiger frühgeschichlicher Siedlungen in Schleswig–Holstein. *Schriften des Naturwissenschaftlichen Vereins für Schleswig–Holstein 28*, 21–41.

Richter, J. 1991: Kainsbakke. Aspects of the Palaeoecology of Neolithic man. *In* Rasmussen, L.W. & Richter, J. (eds): *Kainsbakke. En kystboplads fra Yngre Stenalder*, 72–119. Djursland Museum, Grenå.

Ricklefs, R.E. 2004: A comprehensive framework for global patterns in biodiversity. *Ecology Letters 7*, 1–15.

Rosenlund, K. 1979: Knoglematerialet. *In* Liebgott, N.-K. (ed.): Stakhaven. Arkæologiske Undersøgelser i Senmiddelalderens Dragør. *Nationalmuseets Skrifter. Arkæologisk-Historisk Række Bd XIX*, 166–167.

Rosenlund, K. 1980: Knoglematerialet fra bopladsen Lundby II. *In* Henriksen, B.B. (ed.): Lundby-holmen. *Nordiske Fortidsminder 6*, 128–142.

Schild, R. 1984: Terminal Paleolithic of the North European Plain: a review of lost chances, potentials, and hopes. *Advances in World Archaeology 3*, 193–274.

Sommer, R. & Benecke, N. 2003: Post-Glacial history of the European seal fauna on the basis of sub-fossil records. *Beiträge zur Archäozoologie und Prähistorischen Anthropologie 6*, 16–28.

Sommer, R. & Benecke, N. 2004: Late and Post-Glacial history of the Mustelidae in Europe. *Mammal Review 34*, 249–284.

Sommer, R., Zachos, F.E., Street, M., Jöris, O., Skog, A. & Benecke, N. 2008: Late Quaternary distribution dynamics and phylogeography of the red deer (*Cervus elaphus*) in Europe. *Quaternary Science Reviews 27*, 714–733.

Sondaar, P.Y. 1977: Insularity and its effect on mammal evolution. *In* Hecht, M.K., Goody, P.C. & Hecht, B.M. (eds): *Major Patterns in Vertebrate Evolution*, 671–707. Plenum Press, New York.

Sørensen, S.A. 1996: *Kongemosekulturen i Sydskandinavien*. Jægerspris, Egnsmuseet Færgegården, 192 pp.

Stensager, A.O. 2004: Nyt lys på gammelt fund. *Vendsyssel Nu og Da 23*, 38–43.

Stenstrop, G. 1994: En hval fra stenalderhavet. *Geologisk Nyt 3*, 6–8.

Stewart, J.R. & Lister, A.M. 2001: Cryptic northern refugia and the origins of the modern biota. *TRENDS in Ecology and Evolution 16*, 608–613.

Storå, J. 2001: Reading bones. Stone Age hunters and seals in the Baltic. *Stockholm Studies in Archaeology 21*, 1–86.

Storå, J. & Ericson, P.G.P. 2004: A prehistoric breeding population of harp seals (*Phoca groenlandica*) in the Baltic Sea. *Marine Mammal Science 20*, 115–133.

Storå, J. & Lougas, L. 2005: Human exploitation and history of seals in the Baltic during the late Holocene. *In* Monks, G.G. (ed.): The Exploitation and Cultural Importance of Sea Mammals, *Proceedings of the Ninth Conference of the International Council of Archaeozoology, Durham, August 2002*, 95–106. Oxbow Books, Oxford.

Storch, G. 1992: Local differentiation of faunal change at the Pleistocene–Holocene boundary. *Courier Forschungsinstitut Senckenberg 153*, 135–142.

Street, M. 1997: Faunal succession and human subsistence in the Northern Rhineland 13,000–9,000 BP. *In* Fagnart, J.-P. & Thévenin, A. (eds): *Le tardiglaciaire en Europe du Nord-Ouest*, 545–567. Actes du 119e Congrès national des sociétés historiques et scientifiques, Amiens 1994. CTHS, Paris.

Street, M. 1999: Remains of aurochs (*Bos primigenius*) from the Early Mesolithic site Bedburg-Königshoven (Rhineland, Germany). *Wissenschaftliche Schriften des Neanderthal Museums 1*, 173–194.

Street, M. & Baales, M. 1999: Pleistocene/Holocene changes in the Rhineland fauna in a northwest European context. *In* Benecke, N. (ed.): The holocene history of the European vertebrate fauna. *Archäologie in Eurasien. Band 6*, 9–38.

Svenning, J.-C. & Skov, F. 2005: The relative roles of environment and history as controls of tree species composition and richness in Europe. *Journal of Biogeography 32*, 1019–1033.

Svenning, J.-C. & Skov, F. 2007: Ice age legacies in the geographical distribution of tree species richness in Europe. *Global Ecology and Biogeography 16*, 234–245.

Taberlet, P., Fumagalli, L., Wust-Saucy, A.-G. & Cosson, J.-F. 1998: Comparative phylogeography and postglacial colonization routes in Europe. *Molecular Ecology 7*, 453–464.

Thulin, C.-G. & Flux, J.E. 2003: *Lepus timidus* – Schneehase. *In* Niethammer, J. & Krapp, F. (eds): *Handbuch der Säugetiere Europas. Band 3/II, Hasentiere, Lagomorpha*, 155–185. AULA-Verlag, Wiebelsheim.

Tougaard, S. 2007: Hvalros, *Odobenus rosmarus* (Linnaeus, 1758); Gråsæl, *Halichoerus grypus* (Fabricius, 1791). *In* Baagøe, H.J. & Jensen, T.S. (eds): *Dansk Pattedyratlas*, 244–247 & 258–261. Gyldendal, København.

Troels-Smith, J. 1943: Geologiske dateringer af bopladser i Aamosen. *In* Mathiassen, Th. (ed.): Stenalderbopladser i Aamosen. *Nordiske Fortidsminder III, 3*, 147–164.

Trolle-Lassen, T. 1985: En zooarkæologisk analyse af Ertebøllebopladsen Tybrind Vig, primært baseret på knogler af pelsdyr og kronhjort. Unpublished Master's thesis. University of Aarhus.

Tromnau, G. 1975: Neue Ausgrabungen im Ahrensburger Tunneltal. Ein Beitrag zur Erforschung des Jungpaläolithikums im nordwesteuropäischen Flachland. *Offa-Bücher 33*, 105.

Turner, E. 1990: Middle and Late Pleistocene macrofaunas of the Neuwied Basin Region (Rhineland–Palatinate) of West Germany. *Jahrbuch des Römisch-Germanischen Zentralmuseums Mainz 37*, 133–403.

Udvardy, M.D.F. 1969: *Dynamic Zoogeography with Special Reference to Land Animals*, 445 pp. Van Nostrand Reinhold Company, New York.

Ukkonen, P. 2002: The early history of seals in the northern Baltic. *Annales Zoologici Fennici 39*, 187–207.

Ukkonen, P., Lõugas, L., Zagorska, I., Luksevica, L., Luksevics, E., Daugnora, L. & Jungner, H. 2006: History of the reindeer (*Rangifer tarandus*) in the eastern Baltic region and its implications for the origin and immigration routes of the recent northern European wild reindeer populations. *Boreas 35*, 222–230.

Ukkonen, P., Arppe, L., Houmark-Nielsen, M., Kjær, K.H. & Karhu, J.A. 2007: MIS 3 mammoth remains from Sweden – implications for faunal history, palaeoclimate and glaciation chronology. *Quaternary Science Reviews 26*, 3081–3098.

Van Bree, P.J.H., Vedder, E.J. & T'Hart, L. 1994: Over zadelrobben *Phoca groenlandica* op de kust van continental West-Europe. *Lutra 37*, 97–105.

Van Valen, L.M. 1973: A new evolutionary law. *Evolutionary Theory 1*, 1–30.

Vartanyan, S.L., Garutt, V.E. & Sher, A.V. 1993: Holocene dwarf mammoths from Wrangel Island in the Siberian Arctic. *Nature 362*, 337–340.

Vilhelmsen, H. 2007: Hasselmus, *Muscardinus avellanarius* (Linnaeus, 1758). *In* Baagøe, H.J. & Jensen, T.S. (eds): *Dansk Pattedyratlas*, 164–167. Gyldendal, København.

Wasserzug, R.J., Yang, H., Sepkoski, J.J. & Raup, D.M. 1979: The evolution of body size on islands: a computer simulation. *American Naturalist 114*, 287–295.

Weismann, C. 1931: *Vildets og Jagtens Historie i Danmark*. 564 pp. C. A. Reitzels Forlag, København.

Whittaker, R.J., Willis, K.J. & Field, R. 2001: Scale and species richness: towards a general hierarchical theory of species diversity. *Journal of Biogeography 28*, 453–470.

Willis, K.J., Rudner, E. & Sümegi, P. 2000: The full-glacial forests of central and southeastern Europe. *Quaternary Research 53*, 203–213.

Winge, H. 1899: Om nogle Pattedyr i Danmark. *Videnskabelige Meddelelser fra Dansk Naturhistorisk Forening 51*, 283–316.

Winge, H. 1903: Oversigt over Knoglematerialet fra Mullerupbopladsen. *In* Sarauw, G. (ed.): En Stenalders Boplads i Maglemose ved Mullerup. *Aarbøger for Nordisk Oldkyndighed og Historie 1903*, 194–198.

Winge, H. 1904: Om jordfundne Pattedyr fra Danmark. *Videnskabelige Meddelelser fra Dansk Naturhistorisk Forening 56*, 193–304.

Winge, H. 1908: *Pattedyr. Danmarks Fauna 5. Naturhistorisk Forening*, 248 pp. C. E. C. Gads Forlag, København.

Winge, H. 1919: Levninger af Hvirveldyr fra Stenalders-Bopladsen i Sværdborg Mose. *In* Friis-Johansen K. (ed.): En Boplads fra den ældste Stenalder i Sværdborg Mose. *Aarbøger for Nordisk Oldkyndighed og Historie, Række III, Bind 9*, 128–133.

Winge, H. 1924: Knogler. *In* Broholm, H.C. (ed.): Nye Fund fra den ældste Stenalder, Holmegaard- og Sværdborgfundene. *Aarbøger for Nordisk Oldkyndighed og Historie 1924*, 28–30.

Wojtal, P. 2007: *Zooarchaeological Studies of the Late Pleistocene Sites in Poland*, 189 pp. Institute of Systematics and Evolution of Animals, Kraków.

Wood, R.A., Keen, A.B., Mitchell, J.F. & Gregory, J.M. 1999: Changing spatial structure of the thermohaline circulation in response to atmospheric CO_2 forcing in a climate model. *Nature 399*, 572–575.

Yalden, D. 1999: *The History of British Mammals*, 305 pp. T. & A. D. Poyser, London.